Sound-Based Assistive Technology

Tohru Ifukube

Sound-Based Assistive Technology

Support to Hearing, Speaking and Seeing

 Springer

Tohru Ifukube
Institute of Gerontology
The University of Tokyo
Tokyo
Japan

and

Hokkaido University of Science
Sapporo
Japan

ISBN 978-3-319-47996-5   ISBN 978-3-319-47997-2   (eBook)
DOI 10.1007/978-3-319-47997-2

Library of Congress Control Number: 2016963196

Translation from the Japanese language edition: *Sound-Based Assistive Technology* by Tohru Ifukube, ©
ASJ 1997. All Rights Reserved.

Printed on acid-free paper

This Springer imprint is published by Springer Nature
The registered company is Springer International Publishing AG
The registered company address is: Gewerbestrasse 11, 6330 Cham, Switzerland

Preface

The field of assistive technology did not start attracting attention until recently when people have become more and more aware of the importance of prolonging life, leading to greater expectations in the field of medicine. Recently, however, in addition to technology aimed at prolonging life, people have been placing increasing importance on how to attain good health in order to lead more fulfilling lives. For example, people suffering from even slight disorders are increasingly motivated to regain their health by making use of cutting-edge technology. In light of such expectations, the need for assistive technology has greatly increased, especially in a super-aged society like Japan. Moreover, people who have developed visual or auditory function disorders or communication disorders in their youth must keep living in a society while burdened by a major handicap. Information technology (IT) compensates for weaknesses in human information processing, and the hope has been that IT will serve as a tool to assist those with sensory and/or communication disorders.

Over a period of 45 years, the author has developed a basic research approach for assistive technology, especially for people with hearing, speaking, and seeing disorders. Although some of these tools have practical uses for the disabled, the author has seen how inadequate many of these tools actually are. Moreover, during the course of basic research, the author has experienced how effectively the neuroplasticity of the human body is able to compensate for these disorders. Especially, persons who have developed sensory and/or communication disorders will try to converse through gestures or tactile means, while elderly persons who have developed cognitive impairments will increasingly try to convey something through facial expressions and gestures. However, that is insufficient to properly communicate with others and leads to social isolation, and society is yet to come up with a good approach for how to assist these people.

Under the working title "Sound-Based Assistive Technology," the author would like to show some of the compensation abilities formed by "brain plasticity" and also demonstrate some extraordinary abilities such as the voice imitation skills of the mynah bird and the obstacle-sensing capability of the blind. Furthermore, the author explains the extent to which existing technology could help hearing-, speech-, and sight-impaired people who might all benefit in some way from an enhancement of their extraordinary abilities to recognize and produce speech or to detect sounds in their surroundings. Additionally, it is worth considering how Sound-Based Assistive Technology might be applied to difficulties in areas such as speech recognition, speech synthesis, environmental recognition, virtual reality, and robots.

It is the focus of this researcher to provide an understanding of both the methodology and the basic concepts of assistive technology rather than to give a listing of the entire variety of assistive devices developed in Japan and other countries. In Chap. 1, the author discusses one research approach for assistive technology that is based on the concept of cybernetics, and then introduces a national project called "The Creation of Sciences, Technologies and Systems to Enrich the Lives in the Super-Aged Society" that the author has promoted since 2010. Moreover, the author explains the mechanism of the auditory sense and speech production of human beings that are needed to understand the contents of the book. In Chap. 2, the author explains how auditory characteristics change due to hearing impairment as well as aging, and then introduces some signal processing methods for digital hearing aids. In particular, the author introduces various approaches to signal processing such as noise-reduction methods, consonant-enhancement methods, and speech rate or intonation-conversion methods. Past and recent artificial middle ears are also discussed as one form of hearing aid. In Chap. 3, the author mainly discusses functional electrical stimulation (FES) of the auditory nerves using cochlear implants, especially the history, principle, effects, the basic design concepts and recent progress including auditory brainstem implants. Finally, the author introduces "artificial retina" that has been served by the success of cochlear implants. In Chap. 4, the author introduces tactile stimulation methods, such as tactile aids that support speech recognition of the deaf, vocal training of the deaf-blind, and auditory localization substitutes. The author emphasizes that the findings and technologies obtained from tactile studies will lead to new ideas for tactile displays for virtual reality systems and robots.

In Chap. 5, the author explains recent speech-recognition technologies and discusses captioning systems that convert spoken languages into written characters in real time for the hearing impaired as well as the elderly. Furthermore, to support elderly people with cognitive decline, the author introduces an example of communication robots that remind elderly people to take their medications and of their daily schedule. In Chap. 6, the author introduces an artificial electro-larynx that can produce intonation and fluctuation of the larynx voice, which was based on the vocalization mechanism of a talking bird, the mynah. The author also introduces a voice synthesizer for articulation disorders or speech apraxia, which can produce any speech sounds just by the touching and stroking of the touchpad of a mobile phone. The design of the device was modeled after the vocalization mechanism of a ventriloquist. In Chap. 7, the author mainly discusses two assistive tools for the visually impaired. One of them is a device called the "Tactile Jog-dial" that converts verbal information such as text into speech signals, for which the speech speed can be controlled by blind users, while displaying non-verbal information such as rich text to the tactile sense of a fingertip. The others are mobility aid devices that detect environmental information and display it to the auditory sense of the blind. The mobility aid devices were modeled after the echolocation function of bats and also have the ability of "obstacle sense" by which the blind can recognize obstacles to some extent without any assistive tools.

Finally, the author emphasizes that new research themes concerning the sensory and brain functions of human beings will be provided through the design of assistive technologies and also that new markets will be created in the fields of ICT (information communication technology) such as human interfaces with the brain plasticity function, and IRT (information robot technology), including nursing robots capable of supporting the elderly as well as the disabled.

This book presents a number of different themes, and they are sufficiently independent from one another that the reader may begin at any chapter without experiencing confusion. It should be acknowledged that much of the research quoted in this book was conducted in the author's laboratories both at Hokkaido University and at the University of Tokyo—a fact that could unintentionally result in some bias.

It is the author's fervent wish that this book might at least offer the reader a better understanding of the number of unsolved problems that persist in the field of Sound-Based Assistive Technology. But even more so, if this book serves to increase the number of researchers who are willing to challenge this complex matter, it will have served a useful purpose.

Tokyo/Sapporo, Japan   Tohru Ifukube

Contents

Chapter 1
Basis for Sound-Based Assistive Technology

Abstract In this chapter, the author discusses the methodology as well as the basic concepts of assistive technology based on cybernetics that regards the human body as a feedback system comprising the senses, the brain, and the motor functions. In the latter part of this chapter, the author explains some mechanisms of the auditory sense and speech production of human beings that are needed to understand the contents of the book. The author introduces a national project called "The Creation of Sciences, Technologies and Systems to Enrich the Lives of the Super-aged Society" that he has promoted since 2010.

1.1 Background of Assistive Technology

In the field of medical engineering, medical doctors and engineers have eagerly combined their research efforts with the aim of developing "life support technology" such as artificial organs and regeneration medicine in order to prolong life. However, in the case of people suffering from disorders that cannot be cured by modern medical treatment, there is hope in particular that their disabilities may be compensated for by making use of cutting-edge technology. "Bodily support technology" such as artificial sensors and artificial limbs is a typical example for the compensation of the disabilities. Furthermore, the support technologies will give them the ability to take part in numerous "social participation and activities" for the sake of QOL improvement and job assistance system [1, 2].

Figure 1.1 shows a comparison between the roles of assistive technology and medical engineering, and also shows examples of assistive tools that should be studied and developed [3]. However, the disabilities experienced by people are so diverse and complex that it has been difficult to construct a research approach to develop this assistive technology. At this point, the author would like to explain his personal view regarding the methodology for conducting research in assistive technology. The concept of assistive technology originated from "Cybernetics," published by Norbert Wiener (1894–1964) in 1948 [4]. As indicated in the subtitle "Or Control and Communication in the Animal and the Machine," he described the

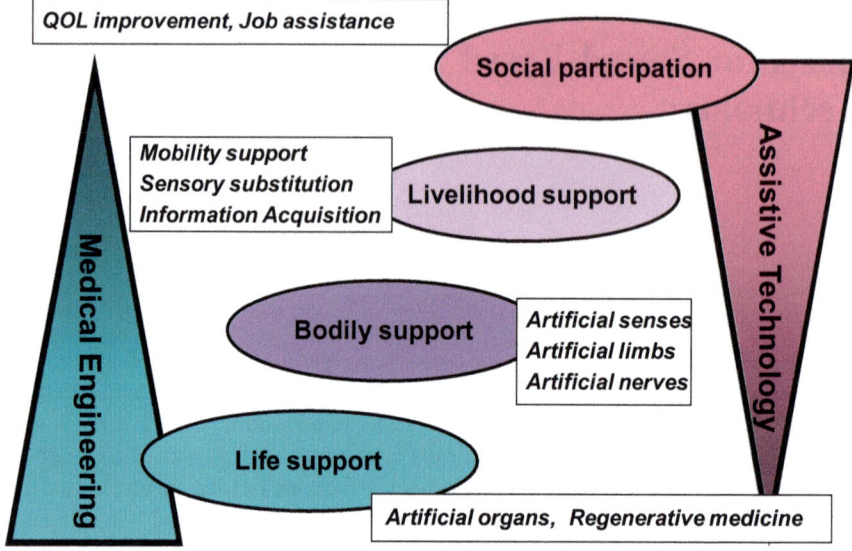

Fig. 1.1 Medical engineering and assistive technology

human functions of senses, brain and limbs—which are essential for living an intellectual life—as having similarities to the functions of the sensors, computers and actuators of automatic machines such as robots. Furthermore, he pointed out that "homeostasis" in both human beings and machines is maintained by feedback control systems. Thus, it can be said that assistive technology research lies in the undercurrent of cybernetics.

As shown in Fig. 1.2, the first point of departure in our research is to investigate and compare the mechanisms of the sensory systems, the brain functions, and motor ability in both healthy and disabled persons based on physiology and

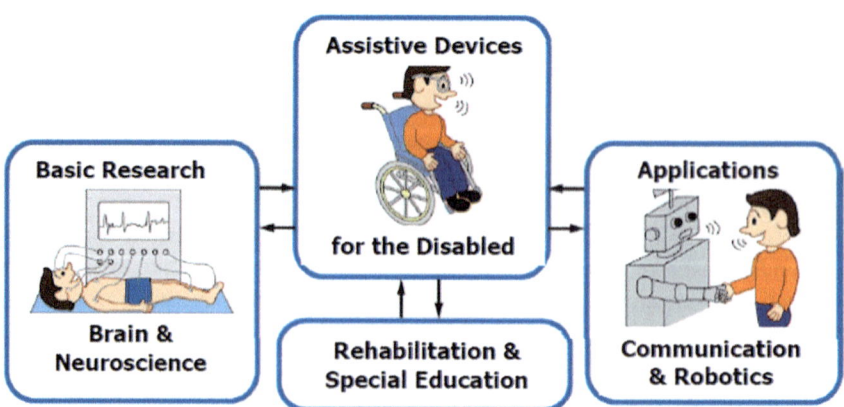

Fig. 1.2 Three stages of assistive technology research as presented in this study

psychophysics, as shown in the left side of the figure. After obtaining knowledge through this fundamental research, the second stage is to design tools that will enable substitution or compensation of a person's senses or limbs, as shown in the center of the figure. If the artificial senses or limbs created for the disabled function well, they can serve as excellent replacements for their original receptors or limbs. Conversely, if the devices are defective or inadequate in some way, it is necessary to come full circle by returning to the original starting point of the first stage. This methodology can thus be characterized by conducting continuous cycles of research that eventually lead to the desired goal.

Furthermore, the artificial senses or limbs thus created can be applied to computer pattern recognition as well as to robot actuators, as shown in the right side of the figure. This final stage could well result in healthy competition among manufacturers that would, in turn, lead to further improvements in this technology, creating a big market and making the price of the tools affordable. The author himself has realized significant progress by utilizing this approach in developing assistive technology over a period of more than 45 years.

On the other hand, as a highly advanced IT society is rapidly approaching, people have become surrounded by all sorts of information tools. As a consequence, more and more people are experiencing distress in their attempts to handle the complexity of both computers and the internet. In particular, a phenomena known as "technostress" is occurring on a wide scale, especially among the less technically inclined. This digital divide resulting from a knowledge gap is, in fact, becoming a major social issue. As information tools are especially difficult for the elderly and disabled to use, machines that can be operated with little or no conscious effort are in high demand. The role of assistive technology is therefore to achieve a state of "universal accessibility" that will empower anyone in society to easily reap the full benefits of the highly advanced IT.

1.1.1 Towards a Super-Aged Society

Currently, more than 25% of the Japanese population are over the age of 65, representing nearly 30 million people, as shown in the left side of Fig. 1.3. As the elderly population increases, so does the number of disabled people. In fact, the ratio of the number of elderly people to the number of disabled people increased from about 31% in 1970 to about 70% in 2011 in Japan, as shown in the right side of the figure [5]. Almost the same situation as that currently prevailing in Japan will extend all over the world, especially in European and East Asian countries. These disabled elderly people all have some form of disability in terms of hearing, speaking, reading, thinking, or moving.

Assistive technology for elderly disabled people is referred to as "geron-technology," which is distinct from "barrier-free design" technology that includes supporting young disabled people. As shown in Fig. 1.4, the barrier-free design for the young uses the "plasticity" of the human body, especially the brain,

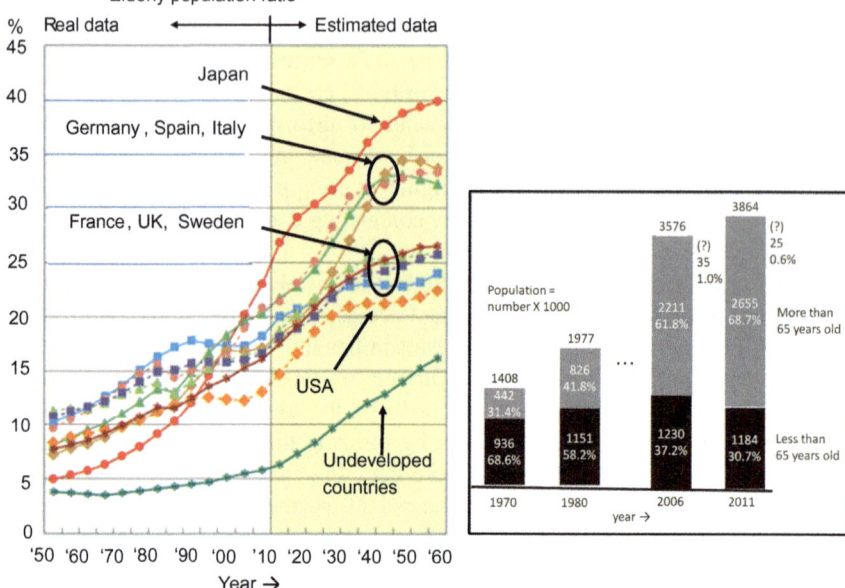

Fig. 1.3 *Left* Elderly population ratio [5], *right* elderly disabled/total disabled in Japan

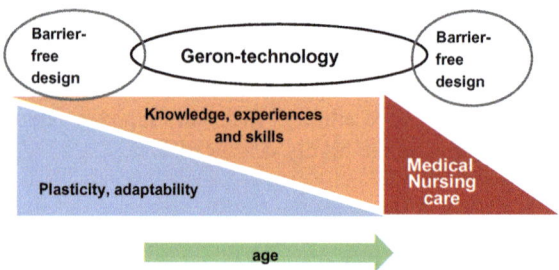

Fig. 1.4 Barrier-free design and geron-technology

because with the help of brain plasticity, residual functions work to compensate for other disordered functions. However, in general, this plasticity function decreases in the elderly, so that they will need to acquire abilities using their "experience" [6]. Overcoming this aging of society is already becoming an increasing burden for the Japanese government.

As a result, many projects have been launched with the Ministry of Economy, Trade and Industry and the Ministry of Labor, Health and Welfare leading the way. The author was asked to promote a new project called "The Creation of Sciences, Technologies and Systems to Enrich Society for the Aged in Japan," which is one of the strategic innovation projects for a super-aged society organized by the Japan Organization for Japan Science and Technology (JST) in 2010. The project started in 2010 and will continue until 2019. It should be possible to apply almost the same

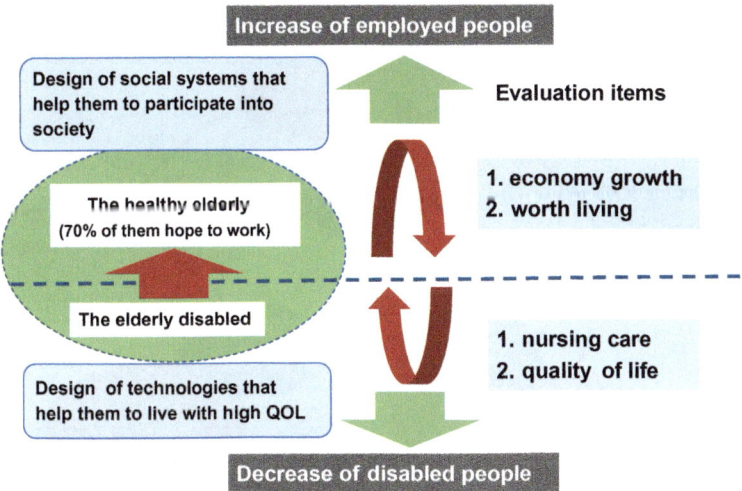

Fig. 1.5 Purpose and evaluation items of national project to enrich both healthy elderly and elderly disabled people

design approach for assistive tools for the elderly disabled as for the young disabled. A secondary purpose of the project is to create new assistive systems that will promote the social participation of healthy elderly persons as well as improve the quality of life (QOL) of elderly disabled persons.

Based on investigations into the functional ability of the elderly, the author divided elderly people into two groups: a healthy elderly group, and a disabled elderly group. In the project, for the healthy elderly, we strongly promote the design of assistive tools and systems so that the healthy elderly can participate in society. For the disabled elderly, we promote the design of rehabilitation technologies and nursing care networks so that the disabled elderly can live with a high QOL. These assistive technologies and systems will decrease social security expenses and may also increase employment and promote economic growth (Fig. 1.5).

Furthermore, five themes were set up concerning information communication technology (ICT) and information robotics technology (IRT) applications in order to realize the aim of our project, as shown in Fig. 1.6. These are:

(a) "Wearable ICT" to assist sensory functions and to detect the physical state for both groups,
(b) "Infra-structured ICT" to design job-matching systems for employment of the healthy elderly,
(c) "Mobility-assistive IRT" to enable free and safe movement within the community as well as at home for both groups,
(d) "Labor-assistive IRT" to help with heavy work and nursing care in daily life for both groups,
(e) "Brain-assistive ICT and IRT" to assist understanding, memory, and information production mainly for the disabled elderly.

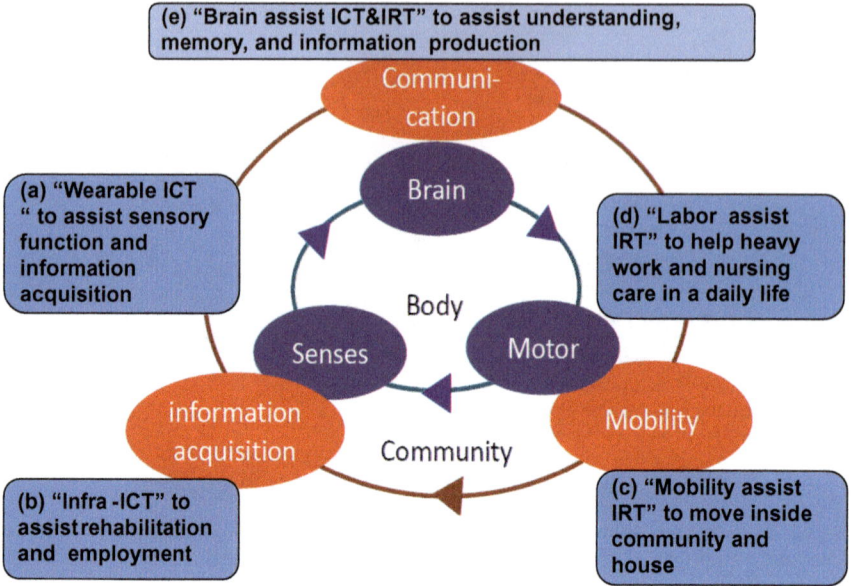

Fig. 1.6 Five themes to assist elderly using ICT and IRT

In the following chapters, the author would like to mention how sounds play an important role in the design of assistive tools in order to support the elderly as well as young people with hearing, speaking and seeing disabilities.

The project is advanced under three stages as shown in Fig. 1.7. The first stage (2011–2013) was to analyze the cognitive and/or behavior functions of the elderly and to investigate needs of the aged society as well as the elderly themselves. The second stage (2014–2016) was to design assistive tools for the elderly and to create systems needed in the aged society. Third stage (2017–2019) is to apply the tools and the systems in a real society and to evaluate the project from a viewpoint of social security, economy growth and QOL of the elderly. These three stages correspond "science", "technology" and "systems" as shown in the figure. The third stage becomes very important to open a big market for mobile phones, automatic driving cars, virtual reality and care robotics that are capable of supporting the elderly as well as the disabled. A project "Labor-assistive IRT" carried out by a joint research of Mitsubishi Electric Engineering (Co., LTD) and Hokkaido University completed in 2014. The author will briefly introduce two projects related "Mobility-assistive IRT" and "Infra-structured ICT". A project related to "Brain-assistive ICT and IRT" will be mentioned in Chap. 5.

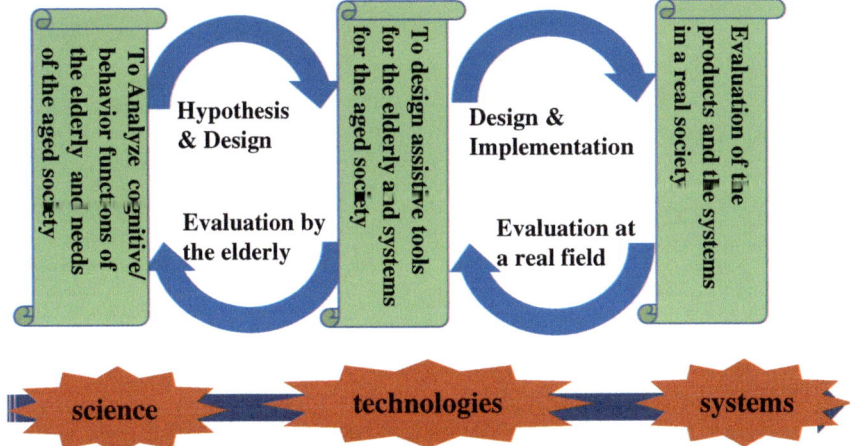

Fig. 1.7 Three stages of a project "The Creation of Sciences, Technologies and Systems to Enrich Society for the Aged in Japan"

1.1.1.1 Autonomous Driving Intelligent Systems to Assist Elderly Drivers

The research project titled "Autonomous driving intelligent systems to assist elderly drivers" was adopted as the mobility-assistive IRT, which has been carried out by a joint research of TOYOTA (Co., LTD) and Tokyo University of Agriculture and Technology. The project has been conducted by Inoue from TOYOTA and his co-researchers. It is reported that traffic accidents dramatically increase in the case of the elderly drivers, as shown in the left of Fig. 1.8. Major accidents are classified as collision accidents between cars, collisions to pedestrians and bicycles, deviation from road lanes and encounter accidents at crossroads as shown in the right of Fig. 1.8. The purpose of the research project is to construct the autonomous driving intelligent system that helps the elderly drivers to avoid these accidents by using an expert driver's model. The expert driver's model works only when there is a risk that an accident will occur. The model was constructed using big data of drive recorders that more than 60,000 professional drivers took just before near misses of car accidents occurred. The risks are predicted based on two kinds of information [7, 8].

As shown in Fig. 1.9, one of them is surrounding information around the car detected by various sensors such as a laser ranging device, several cameras and etc. The other geographic-traffic information obtained by map data, weather reports, GPS and etc. When the probability of the predicted risk is greater than the threshold, the expert driver's model controls a break and a steering wheel instead of the elderly drivers.

They have been also investigating acceptability of the aged society as well as the elderly themselves in order to make the regulations to accept the new technologies.

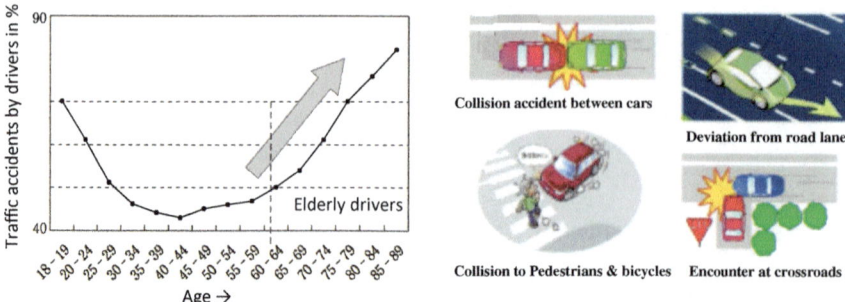

Fig. 1.8 *Left* Traffic accidents versus age of drivers, *right* four major traffic accidents

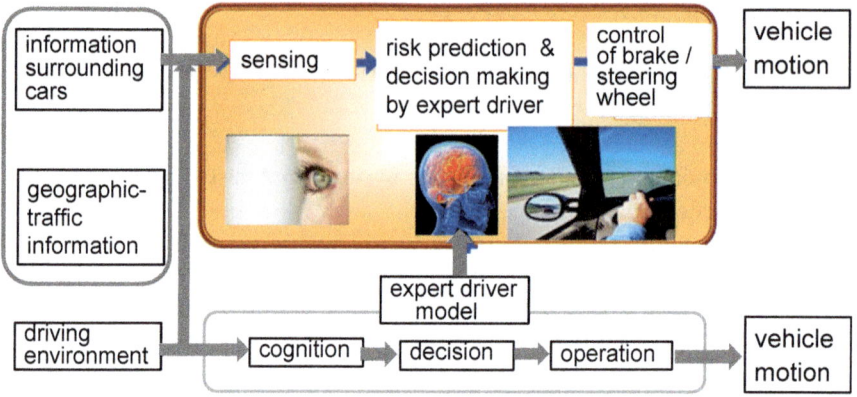

Fig. 1.9 Autonomous driving intelligent system consisting of information detected sensors, geographic-traffic information and expert driver's model

It is expected that the autonomous driving intelligent systems may lead to expand their actions range and prompt social participation.

1.1.1.2   "Senior Cloud" Using Knowledge, Experiences and Skills of the Elderly

The research project titled "Senior cloud using knowledge, experiences and skills of the elderly" was adopted as the infra-ICT, which has been carried out by a joint research of the university of Tokyo and Japan IBM. The project has been conducted by Hirose at the university of Tokyo and his co-researchers. As shown in the left of Fig. 1.10, recently diversify of working style of the elderly has rapidly increased. For example, their working time has shifted from full-time to any-time, and their working place is from wide area to their home. Furthermore, their acquired knowledge, experiences and skills also have diversified [9, 10].

Fig. 1.10 *Left* Working styles of the elderly, *right* an example of mosaic model (skill mosaic)

The purpose of the research project is to promote working and participations in society by constructing "MOSAIC model" and job-matching platform that positively utilize the diversity and integrate their acquired abilities. The concept of the mosaic model is shown in the right of Fig. 1.10. For example, if a senior A has 80% of skills of averaged labor in the working field X, a senior B has 10%, a senior C has 10%, then one virtual worker with 100% skill is expected by integrating these three persons' skills. As well as the skill mosaic model, it can be constructed that a working time mosaic model, a working place mosaic model, an experience mosaic model and so on.

Two interfaces and one platform were designed in order to realize the concept of the senior cloud; knowledge acquisition interface, knowledge structuring platform and knowledge transfer interface as show in Fig. 1.11. The knowledge acquisition interface includes "Question first method" that may elicit the various ability and characteristics from interviews and communication to them. As a result, the interface can estimate who, when, what, where, how the elderly can do. In the knowledge structuring platform, various mosaic models are constructed based on the data obtained by the knowledge acquisition interface and job matching for the elderly are performed. In the knowledge transfer interface, ICT and IRT such as virtual reality technologies and tele-existence robots are used so that the elderly may transfer their knowledges, experiences and skills to the people who work together in the desired location and favorite time. It is expected that the senior cloud systems may lead to sustain their purpose in life and also maintain their healthy as well as increasing their QOL.

1.1.1.3 Future Prediction Expected from Assistive Technologies

Furthermore, our research approach based assistive technology using the ICT and IRT may contribute to increase worker population, to decrease social security expenses and to open new markets. As population rate of the elderly is increasing in the other countries, it is expected that the assistive technology as well as the

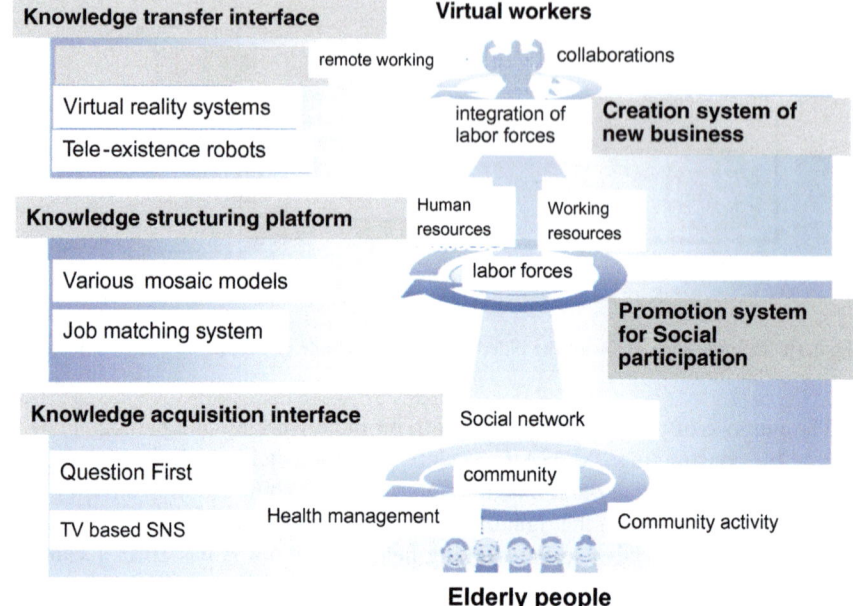

Fig. 1.11 A concept of "Senior cloud" consisting of knowledge acquisition, knowledge structure and knowledge transfer

geron-technology will take an important role for almost all countries. As a result, our national project may contribute to create new export industries and also to increase the economy growth in the near future as shown in Fig. 1.12.

Under this background, the assistive technology to support hearing, speaking and seeing will occupy an increasingly important position, especially for prompting a social participation of the elderly as well as the disabled. The author would like to focus on the sound-based science and technology because they are essential for communication and information acquisition in highly-sophisticated information society.

Fig. 1.12 Assistive technology for the elderly as well as the disabled promotes both economy growth and increase of QOL, making new industries

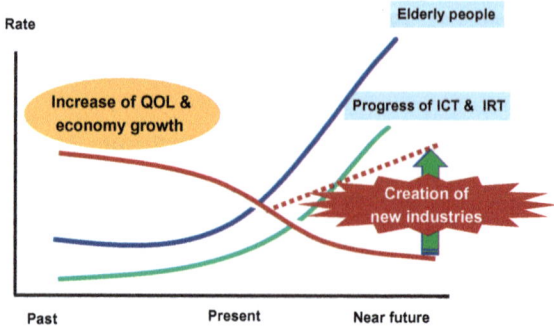

1.2 Roles of Sound in Assistive Technology

Sound became a vehicle to transmit information over a great distance ever since the Earth began to have an atmosphere. In particular, while both animals and humans have been using sound for communication, the invention of speech has enabled mankind to create civilizations and cultures. In this regard, it can be said that speech serves as a social tool for the transmission of civilizations and cultures. In fact, although some tribes do not possess any written language, there is no civilization that exists that lacks a spoken language.

On a different note, when speech is considered on an individual basis, it could be categorized as a tool for thought. Specifically, the acquisition of speech has enabled humanity to make full use of word combinations that give rise to abstract thought, which, in turn, leads to creativity. Such silent speech, which takes place in the head, is known as "internal speech". Thanks to it, the cerebral cortex of human beings has acquired increasingly complex capabilities. It gives one pause to consider that this extraordinary process arose out of the mere existence of air and its ability to transmit information through the vehicle of sound.

The well-known Helen Keller (1880–1968) (see Photograph 1.1) was once asked whether it was a greater disability to be blind or deaf. She answered: "To be deaf is to lose the tool of sound—a tool that greatly enhances one's ability to be socially active as well as to think. In that sense, being deaf is more inconvenient than being blind." Moreover, as speech, derived from sound, provides the building blocks that lead to the creation of abstract thought in conjunction with cerebral development, speech concepts could still be formed without the use of sight. In contrast, for those with congenital deafness, the lack of speech will undoubtedly retard development of the cerebrum. In severe cases, this could result in little or no cerebral development beyond a mental age of around nine.

It goes without saying that the transmission of speech is an important object of study in the field of engineering. In this regard, when two or more people are in close proximity, communication can be accomplished through the vehicle of air; however, when separated by a great distance, other media become necessary if the message is to be received. This was the essential reason the telephone was invented. In particular, it is worth noting that Alexander Graham Bell (1847–1922), the inventor of the telephone, was a linguist whose wife was hearing impaired. His invention was therefore motivated by a fervent desire to communicate with his wife by any means possible. With this in mind, it is no exaggeration to say that an extraordinary effort by one man to assist his wife is what eventually lead to the development of today's enormous telephone system.

When information is conveyed over a great distance, as in the case with telephones, radios and television, one must consider both the transmission of sound as well as its reproduction on the receiving end. Specifically, in order to understand how transmission takes place, matters such as information processing, compression, and restoration must be analyzed from an engineering perspective.

Photograph 1.1 Alexander
Graham Bell with Helen
Keller and Annie Sullivan at
the meeting of the American
Association to Promote the
Teaching of Speech to the
Deaf, July 1894, in
Chautauqua, N.Y.

Another application of engineering is technology allowing communication between human beings and machines. There was an article in a newspaper written in 1901 [11] that attempted to predict future technology over the next century. In particular, the authors foresaw great technological progress in the field of communication. However, they believed that technology would eventually make it possible for humans to communicate with animals. Despite this prophecy's obvious inaccuracy, it was not entirely off the mark if one were to replace the word "animals" with "machines." Devices that can understand as well as respond to human speech are being applied to automatic recognition and sound synthesis. If the quality of such devices is sufficiently improved, they could provide adequate substitutes for the hearing and speech impaired.

Although speech synthesizers have been successfully produced, sound-recognition systems require further research. More specifically, a more in-depth investigation must be conducted regarding neurological functions that pertain to the understanding and production of speech. Due to the lack of such neurological data, it is difficult to determine the best way to equip computers with speech capability. As various biomedical measurements have recently become available for brain analysis, it might soon be possible to finally understand the brain —by far the most mysterious part of human beings. Furthermore, as our understanding deepens, we will be in a better position to assist the hearing and speech impaired. In any case, no real progress in sound-recognition technology will be possible without such a clearer understanding of the brain.

This chapter will first outline the process of how sound is received and how this information is conveyed to the brain through the auditory nervous system. In doing so, some consideration will also be made as to why speech disorders occur.

1.3   Structure and Functions of the Auditory System

1.3.1   External Ears

Figure 1.13 shows the structure of the hearing organs from the auricle to the internal ear. The external auditory canal is a winding tube 10 mm in diameter and 35 mm in length with a tympanic membrane obliquely placed at the bottom. Acoustically speaking, the external auditory canal serves as a resonance tube and has the effect of increasing the decibel level from 10 to 20 dB for sound in the range of 2000–4000 Hz, as illustrated in Fig. 1.14.

However, the external ear, including the auricle, does not merely collect and filter sound. For example, when listened to through headphones, sound is heard as if it were originating from inside the head; whereas, in the case of the external ear, the feeling is that the sound is coming from outside. Another important function of the auricle is to provide the listener with a sense of whether the sound originates from above or below. As will be discussed in Chap. 7, when the auricle is covered with a substance such as silicone, a sound coming from below the listener's ear is actually heard as if coming from above.

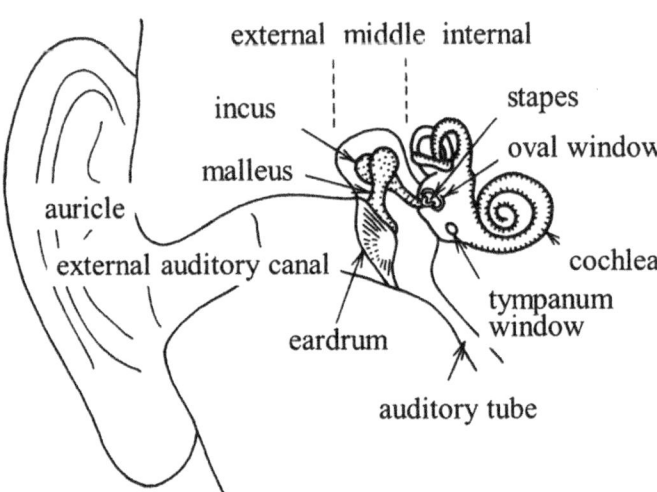

Fig. 1.13 Schematic representation of auditory organs

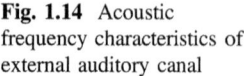

Fig. 1.14 Acoustic
frequency characteristics of
external auditory canal

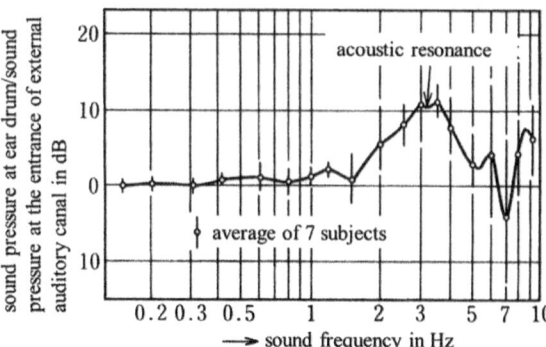

When listening through headphones, a sound recorded through a microphone that has been placed on a tympanic membrane is experienced as originating from outside of the head—a phenomenon known as external sound localization. More specifically, this artificially produced sound is made possible by first using a computer to calculate the head-related transfer function (HRTF) between the origin of the outside sound and the tympanic membrane. In the field of virtual reality (VR) —also called artificial reality—NASA has succeeded in creating external sound localization that offers a very realistic presence by calculating the HRTF using a digital signal processor in real time [12].

A further function of external ears is illustrated in the case of people who are completely blind, most of whom make use of their external ears to develop what is known as an "obstacle sense"—the ability to recognize obstacles by depending exclusively on sound.

1.3.2   Middle Ear

It is said that the origin of the hearing receptors of mammals was a sensor in fish found along its sides, called a "lateral organ". It is a sensor that measures the speed of its own movement and also detects the speed of the seawater when a fish is stationary. Under magnification in a microscope, it is found that the lateral organ is lined with many "hair cell receptors" (see Fig. 1.20a). As will be explained later, these hair cells play an important role as auditory receptors.

As fish evolved into mammals and started living on land, the middle ear developed in such a way that the highly sensitive hair cells could be used to detect air movements. When a sound wave is emitted toward seawater, only about 0.1% of the sound's energy enters the water from the air; the rest is reflected when striking the surface of the sea. It is due to this phenomenon that hair cells in the lymph fluid cannot be used efficiently. Therefore, in order for a sound wave to enter the lymph fluid effectively, acoustic impedance matching by the middle ear became necessary.

Gradually, through evolution, the large tympanic membrane became capable of receiving sound waves. Once this occurred, it was possible for this membrane to then transmit the vibration to a small bone called "the hammer" that is one of auditory ossicle, finally reaching another small bone known as "the stapes" that is located on the entrance of the cochlea called oval window (see Fig. 1.15). The ratio of the area of the tympanic membrane to that of the oval window is about 20 to 1. This great difference in area results in the compression of the sound energy received at the bottom of the stapes bone. This compression increases the sound pressure per unit area by about 26 dB. The auditory ossicle muscle also contributes to the acoustic impedance by functioning like a kind of seesaw. Therefore, it is the middle ear that converts the acoustic impedance.

Furthermore, the auditory ossicle muscle functions to prevent strong sounds from entering the internal ear by lowering the sensitivity of the middle ear through feedback. As this muscle is also quite delicate, certain problems such as middle ear inflammation can easily occur. Therefore, as will be explained in a later chapter, artificial middle ears have been developed to substitute for the functions of natural middle ears.

By making a rough estimate of the amplitude of the vibration of the tympanic and basilar membranes from the auditory threshold, it is found that the amplitude of the vibration of the tympanic membrane is roughly equal to the diameter of a hydrogen atom. In the case of the basilar membrane, the amplitude of the vibration is roughly the size of the nucleus of a hydrogen atom. 16,000 hair cells aligned in four rows at the basilar membrane detect subtle vibrations. If mammals had evolved on land from the beginning, they would have bypassed the rather roundabout, and largely impractical, evolutionary process just described. Instead, they would have quickly developed sharp sensors to detect air vibrations, much like microphones do today. Therefore, it can be said that the rather high incidence of hearing disorders in mammals arises in part as a result of a cumbersome evolutionary process.

1.3.3 Mechanism of Frequency Analysis in the Internal Ear

After mammals started living on land, hair cells were aligned on a thin "basilar membrane" inside a compact, snail-shaped cochlea. The basilar membrane is covered with lymph fluid that contains elements similar to seawater. The reason for the cochlea's snail-like formation was to enable it to wind itself roughly $2^3/_4$ times into a compact coil. If this coil were completely stretched out, its function would be like a rather long tube, reaching a length of 35 mm (Fig. 1.15). Additionally, while the cochlea gradually becomes narrower as it approaches the end tip, the basilar membrane, in contrast, gradually becomes wider towards the tip. In particular, the width of the basilar membrane that contains the hair cells is 0.1 mm at the entrance but 0.5 mm at the very end.

The pressure applied to the stapes and oval window, located at the entrance of the cochlea, reaches the cochlear tip via the lymph fluid. This pressure, in turn,

Fig. 1.15 Schematic representation of extended cochlea

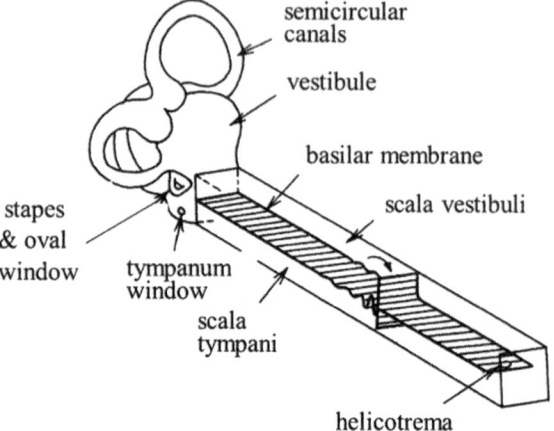

bends the basilar membrane. This bending, appearing at both the oval window and the round window (tympanum window), reaches the tip within about 5 ms. Some hypotheses that frequency analysis occurs at the basilar membrane have existed for a long time.

Among these, Helmholz's (Hermann Ludwig Ferdinand von Helmholtz, 1821–1894) hypothesis gained widespread support for some time. He believed that the thin fiber on the basilar membrane played a role, similar to the keys on an automatic piano, in allocating sound according to its pitch. That is to say, when a certain sound enters the ear, only specific keys vibrate, meaning that an extremely sharp frequency analysis is already done at the basilar membrane. As will be explained more fully later, this hypothesis has been proved from a different point of view.

On the other hand, Békésy (Georg von Békésy, 1809–1972) doubted that only the thin fiber vibrated as the basilar membrane stretched itself out in a wide manner. Therefore, he used elephants' ears, which are much larger than those of humans, to observe the vibration of the basilar membrane. As shown in the left side of Fig. 1.16,

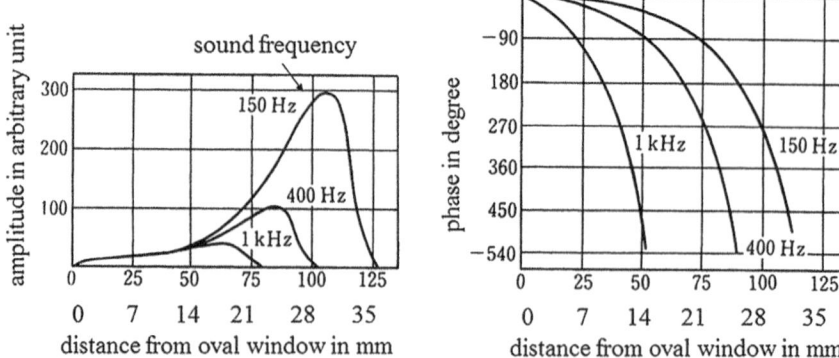

Fig. 1.16 Amplitude and phase of vibratory pattern of basilar membrane

the results indicated that the vibration of the basilar membrane extends over a very large area, and that it disappears at the end of the cochlea [13]. This is a sort of traveling wave, corresponding to the waves that disappear at the end of a long rope fixed to a wall after being shaken by a hand. For this reason, Békésy's theory is known as the "traveling wave theory."

This means that the part that vibrates most depends on the frequency of the sound. Specifically, a broad frequency analysis occurs because high-frequency sounds are analyzed at the entrance of the cochlea, whereas low-frequency sounds are analyzed at the end of the cochlea. The average speed of the traveling wave is 6–7 m/s. As the wave approaches the end of the basilar membrane, it slows down because the basilar membrane gradually softens towards the end. Békésy also performed detailed research on the relation between the intensity and the phase of the vibration, thereby clarifying that the phase moves in a negative direction from the part where the vibration is largest as shown in the right side figure. This means that there is a delay factor in the transmission function. Although this place theory was later revised, the fundamental idea is still in effect.

1.3.4 Hair Cells and the Auditory Nerve System

According to Békésy's study, the sharpness in the resonance of the basilar membrane can be represented by a Q-factor (center frequency/3-dB bandwidth) of 6. Since this value corresponds to a slope (sound intensity/frequency) of 6-dB/octave in a low-frequency region and -20 dB in the high-frequency region, the resonance characteristics of the basilar membrane do not show such a high resolution.

The human auditory system has an extraordinary high-frequency resolution for sounds, making it possible to differentiate 1003 Hz from 1000 Hz. This fact means that the differential limen of the sound frequency is only 0.3 ((1003–1000 Hz)/1000 Hz). Békésy hypothesized that this high-frequency resolution occurs in the auditory nervous system. Katsuki attempted to prove that the auditory nervous system might sharpen the low-frequency resolution. He did this by using a nervous lateral inhibition function predicted by his studies [14]. Those studies shall be explained in detail later.

Békésy indirectly proved the existence of such a sharpening mechanism by utilizing the human tactile sense. Specifically, he designed a membrane on a tube filled with liquid water, similar to a basilar membrane on a cochlea filled with lymph liquid inside an ear, as shown in Fig. 1.17. He then attached the membrane to a human forearm in order to correspond hair cells and the auditory nervous system to tactile receptors and the tactile nervous system, respectively.

Although the membrane on the tube vibrated widely by applying a vibration to the edge of the tube, the perceived vibratory patterns were much sharper than the vibrating membrane patterns, as schematically represented in Fig. 1.18. From this simulation of the cochlea using the tactile sense, Békésy indirectly proved that sharpening in the auditory sense might be realized by a function of the nervous system. He further hypothesized that the same function in the nervous system as the

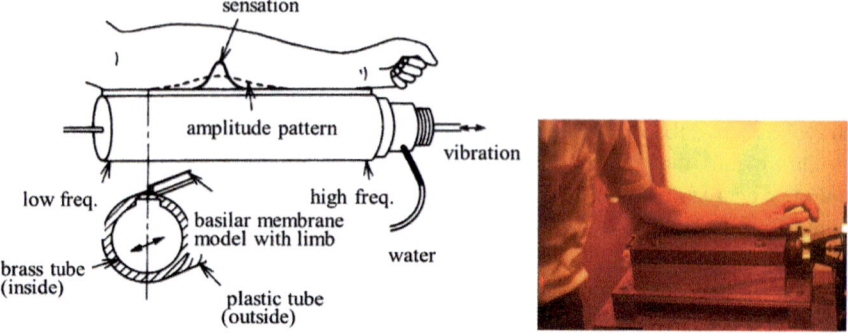

Fig. 1.17 Fluid tube model (*left*) of cochlea and the photograph (*right*)

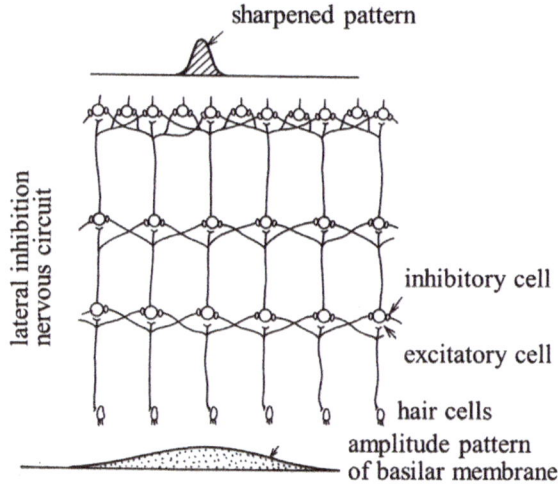

Fig. 1.18 Sharpening characteristics inside auditory nervous system

auditory sense also exists in the tactile nervous system. This idea was the origin of a "tactile aid" that utilizes the tactile sense of the hearing impaired as a substitute for the auditory system.

However, there were inherent limitations to Békésy's experiments, as mentioned in the previous paragraph. Since he only used cochlea removed from elephant cadavers, measurements of basilar membrane movement could only be made under intense pressure. Specifically, it has long been a supposition that a higher frequency resolution might be observed in the cochlea membrane even for weak sounds if a human cochlea were used.

Since a new method of measurement, known as the Mössbauer method [15], was discovered in 1980 to detect living cochlea membrane movements, the frequency-analyzing mechanism in the cochlea has been revised by many auditory

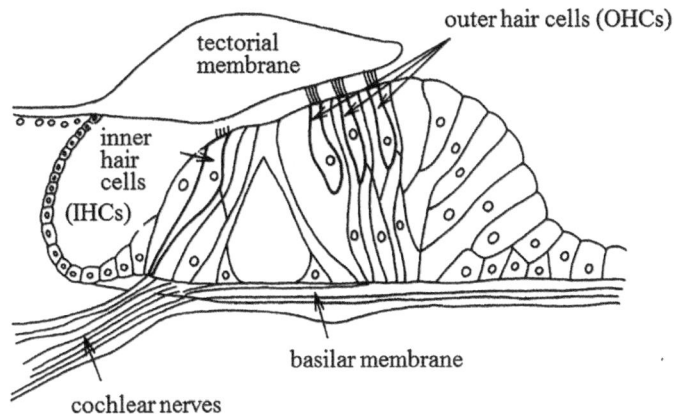

Fig. 1.19 Structure of Corti organ, and OHCs and IHCs

neuroscientists. According to the findings obtained through the Mössbauer method, it was learned that when the stimulating sound level is less than 30, the living cochlea has a very high resonance level (Q): approximately 13 in the low-frequency region and 100 in the high-frequency region.

Furthermore, it has been proven that the basilar membrane itself has such a high frequency resolution and the frequency analysis works adaptively according to sound level due to feedback control of the hair cells, as shown later. The basilar membrane works as a low-frequency analyzer for loud sounds and as a high-frequency analyzer for weak sounds.

As mentioned, the extraordinarily weak movements of the basilar membrane are received by 16,000 hair cells, arranged in four rows on the membrane. The hair cells inside the cochlea of human beings are divided into two types: outer hair cells (OHCs) and inner hair cells (IHCs) as shown in Fig. 1.19. The total number of OHCs, arranged in three rows, is around 12,000, whereas the total number of IHCs, arranged in one row, is about 3500. One OHC is about 8 μm in diameter and has around 140 hairs, whereas an IHC is about 12 μm in diameter and has around 40 hairs. The hair cells are so small and delicate that their function decays through natural aging and is sometimes lost due to the influence of certain antibiotic medicines. In fact, some forms of hearing impairment are caused by the loss of function of the hair cells.

Although all hair cells are inside the Corti organ (see Fig. 1.19), Corti discovered that only the hairs of the OHCs are connected to a tectorial membrane, causing them to bend and vibrate according to the basilar membrane's movements. Although the hairs of the IHCs are not connected to the tectorial membrane, the hairs are bent through the movement of lymph liquid inside the cochlea when the vibration of OHCs increases beyond a certain intensity. It is said that the frequency resolution in the basilar membrane is caused by a combination of OHCs and IHCs.

Through the bending of these hair cells, their mechanical energy is converted into electrical energy, thereby increasing the receptor potential of the hair cell. The

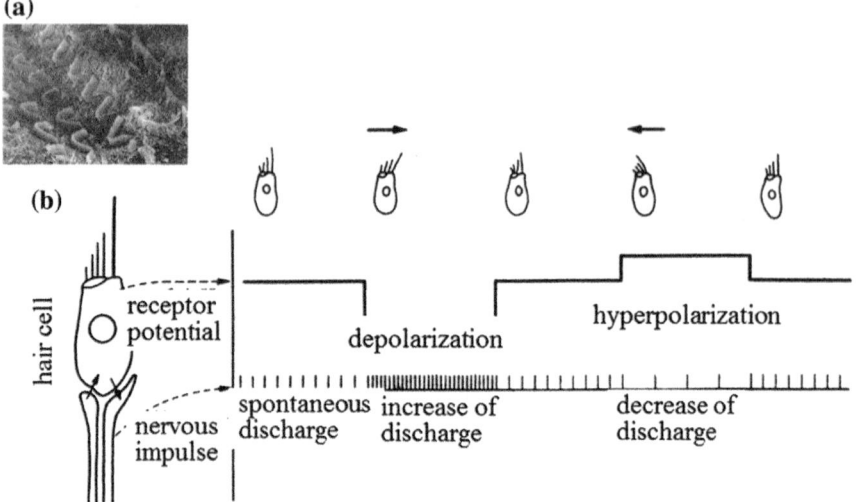

Fig. 1.20 **a** Micrograph of hair cells inside cochlear of guinea pig, **b** auditory nervous impulses produced by bending hairs of hair cells

increase in the potential enables neurotransmitters to travel to the cell membrane, where they are released outside the cell. The transmitters raise the synaptic membrane potential of auditory nerves, resulting in an increase in the membrane potential as shown in Fig. 1.20. When the membrane potential exceeds a certain threshold, impulses are produced that travel to the central nervous system (CNS) through auditory nerves.

The nerve fibers through which the impulses travel from the peripheral nervous system to the CNS are called "afferent nerves," but when the impulses travel from the CNS to the peripheral nervous system, they are called "efferent nerves." Since all receptors are connected by both kinds of nerves, the receptor potential undergoes complex changes through control by such nerves.

Before the roles of the OHCs and the IHCs were clarified, Katsuki had proved the existence of the sharpening mechanism of the auditory frequency resolution from the viewpoint of neuroscience. Specifically, he was able to measure the impulses of auditory neurons in cats by utilizing a microelectrode technique while producing sound stimulation of various intensities and frequencies. As the auditory nervous system of humans is basically the same as that of cats, as schematically shown in Fig. 1.21, it was believed that Katsuki's experimental results could also be applied to understanding the human auditory system.

Auditory nerves that are synaptically connected to hair cells are called "auditory primary neurons." In Katsuki's study, after inserting a microelectrode into one of the auditory primary neurons, cats were subjected to sound stimulation of varying intensities and frequencies in order to measure the thresholds at which the neurons fired. By plotting the threshold as a function of the sound frequency, it became clear

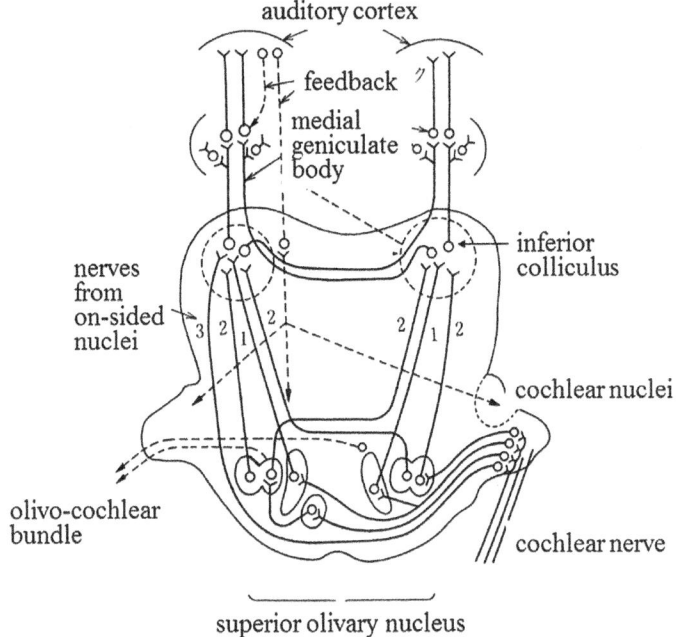

Fig. 1.21 Schematic representation of auditory nervous system

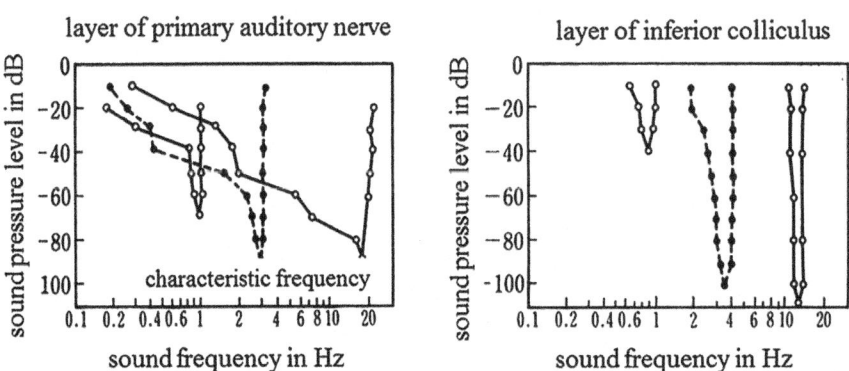

Fig. 1.22 Sharpening characteristics inside auditory nervous system

that, at a certain frequency, the threshold showed a minimum. This frequency is known as the "characteristic frequency" (CF). Katsuki's study also indicated that the threshold gradually increases at lower frequencies than the CF, while it does not increase so gradually at higher frequencies than the CF, as shown in the left part of Fig. 1.22. It is assumed that the pattern showing the frequency-dependent threshold would faithfully reflect the spatial movement pattern of the basilar membrane. This similarity implies that the frequency resolution at the level of the auditory primary system is also broad, like that for the basilar membrane.

Most nerve fibers connecting to the auditory primary neurons cross over at the medulla level, called the "olive nucleus," and then reach the next layer, known as the "inferior colliculus" (IC). By inserting a microelectrode into one of the IC neurons, the thresholds for sound stimulation were obtained as a function of the sound frequency, using the same measurement method as for primary neurons. The results are shown in the right part of Fig. 1.22. Comparing the two graphs, it is apparent that the threshold-frequency pattern for the IC neurons is much sharper than that for the primary neurons. In support of Békésy's hypothesis, this fact indicates that a sharpening mechanism does indeed exist between the primary neurons and the IC neurons.

It is hypothesized that the "lateral inhibition" neural circuit performs an important function in the sharpening process. Many nerve fibers that horizontally connect to surrounding neurons have been found in the IC layer. Furthermore, it is known that nerve cells include a number of inhibitory neurons that decrease the membrane potential of the surrounding neurons by means of the lateral inhibitory function. In this way, the weakly firing inhibitory neurons cause a slight decrease in the membrane potential of the surrounding neurons. Consequently, the frequently firing neurons fire at an even higher rate than the surrounding neurons. This process causes the broad threshold-frequency pattern in the primary neuron layer to be sharpen in the IC neuron layer.

It is assumed that the sharpening mechanism functions at least for a sound intensity greater than 30 dB, an intensity for which the sharpening mechanism of the basilar membrane ceases to function. Actually, recent studies of cochlear implants (see Chap. 3) still indicate that the auditory nervous system plays an important sharpening role, as patients with such implants retain this sharpening ability despite the failure of their basilar membranes to function.

Békésy hypothesized that the visual and tactile senses possess the same sharpening function as the auditory sense. Based on psychophysical experiments, he verified that the sharpening function is also present for the visual and tactile senses. He coined the term "sensory inhibition" to indicate the sharpening function that can be observed by other senses. He gave the name "neural unit" to the spatial band-pass filter. A method by which to estimate the neural unit will be discussed later in this chapter.

It has been found that the threshold-frequency pattern becomes sharper at a higher level in the auditory nervous system. This pattern becomes sharpest at the next layer of the IC, called the "medial geniculate body." Furthermore, the sensation of loudness with regard to sound intensity is also sensed in the medial geniculate body, since the firing frequency of impulses is linearly proportional to the degree of loudness within the level of the medial geniculate body. It has been ascertained that the cerebral cortex processes the more complex sounds, such as their tone color and time sequence.

1.4 Mechanism of Speech Production

1.4.1 Structure of the Speech Organ

There are many animals that can produce vocal sounds. The majority of these animals can express their emotions or thoughts by changing the tone of the sounds they make. However, even in the case of chimpanzees, it is nonetheless impossible for them to produce sounds like human speech. This is because their speech cortex does not function like the "Broker's area" in humans. Specifically, the Broker's area functions in a way that is superior to the speech cortex of other animals. Thereby, the human speech organs—including the lips, tongue, and jaw—constitute the essential differences between humans and other animals.

As shown in Fig. 1.23, the speech-production process is divided into three stages. The first stage involves exhaling through the larynx; the second is characterized by the glottal sounds produced in the vocal folds, and in the third stage, the final sounds are formed through the use of the vocal tract—including the tongue, lips and jaw.

By inhaling during the first stage, as the lungs and chest expand, they produce the elastic force necessary for the exhalation pressure. As people exhale, they adjust the pressure to produce specific sounds by consciously controlling the exhalation muscle. By increasing the exhalation pressure, both the intensity and pitch

Fig. 1.23 Three stages of speech production

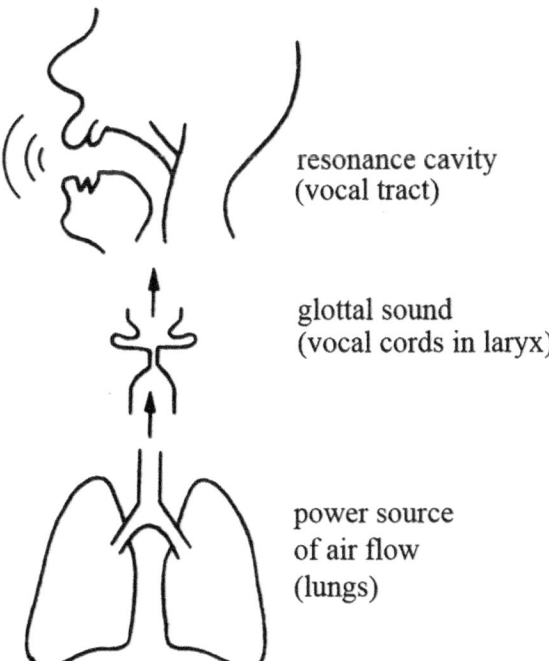

resonance cavity
(vocal tract)

glottal sound
(vocal cords in laryx)

power source
of air flow
(lungs)

frequency of the sounds increase. This positive correlation between the intensity and the pitch frequency makes it possible to change the pitch frequency just by controlling the exhalation pressure. One example of this is in the case of people who have lost their larynx yet retain control of pitch frequency by the use of exhalation pressure, as shown in Chap. 6.

The exhalation pressure that is produced during the first stage becomes the power source for speech in preparation for the second stage. Although the exhalation pressure is held relatively steady at a frequency of around 1 Hz, it is converted into an alternative form of acoustic energy at a pitch frequency of around 200 Hz through the larynx. The name "larynx" can be traced to the Greek word "larugx (λαρνγξ)," meaning to shout or the Latin word "lurcare," meaning to eat voraciously. As a result, the meaning of larynx in modern English has been used to imply either voice or swallowing.

Figure 1.24 schematically represents a structure of the speech organ. From the figure, it is found that the larynx is located under the pharynx, where the trachea and the esophagus separate. The original role of the larynx was to protect the trachea from food toxins. Its structure consists of some cartilages with attached muscles and vocal folds that produce sounds (see Fig. 1.25a). Muscle ligaments and mucous membranes are attached to the bones and vocal folds that produce sounds. The vocal folds' elastic tissue is composed of muscle ligaments and mucous membranes, which are located inside the cartilage, as shown in Fig. 1.25b.

The tension and elasticity of the vocal folds are changeable, as is the length and width of the vocal folds themselves. Furthermore, the open space within the vocal folds is adjustable and the location of the vocal folds themselves can be moved vertically through movement of the larynx. While a person is speaking, the above changes are constantly occurring. Through the evolution of the larynx from a simple valve to a speech-production organ, the larynx has developed complex movements by controlling numerous muscles that work together. In the event of a disease such as cancer of the larynx, a surgeon will remove both those muscles as well as the larynx itself. Following such an operation, information to control the

Fig. 1.24 Schematic representation of speech organ

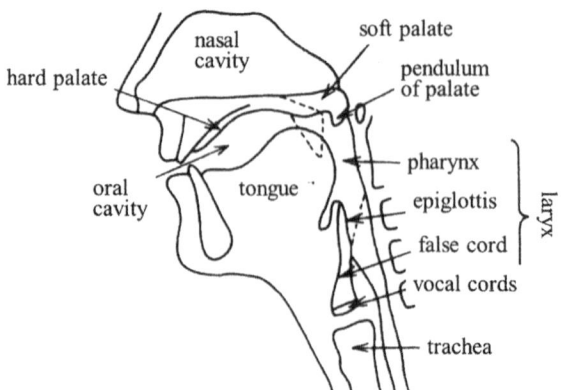

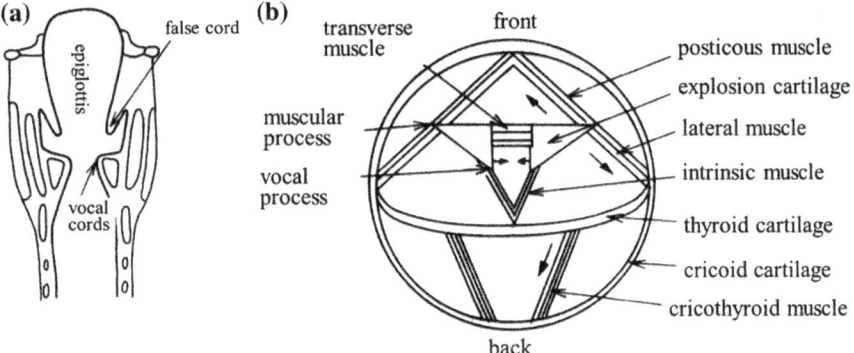

Fig. 1.25 **a** Profile of larynx, **b** schematic representation of laryngeal muscle

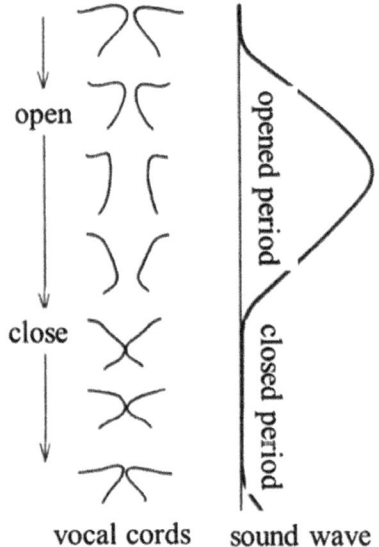

Fig. 1.26 Movement of vocal folds and glottal sound wave

larynx is lost. Therefore, it becomes difficult to produce natural speech, even by using such substitute speech mechanisms as an artificial larynx or through "esophageal speech."

In the process of producing glottal sounds, first the vocal folds bend inwardly and then the glottis closes, as shown in the left figure of Fig. 1.26. Next, the pressure under the glottis increases, making the glottis open. When the glottis is open, air is able to flow through. The effect of this air flow causes the glottis to close because of both the effect of Bernoulli's principle and the elasticity of the vocal folds. As this opening-closing process repeats, the vocal folds vibrate, thereby resulting in the production of glottal sounds. The vibratory frequency of the vocal folds is determined by the elasticity, tension and mass of the folds. Each of these adjustments is performed by the muscles attached to the cartilage of the larynx.

The frequency of the glottal sound is mainly adjusted by controlling the tension and the mass of the vocal folds. The adjustments are performed by the vocal fold muscle and the cartilage muscle inside the larynx. Very low frequency speech sounds, known colloquially as "thick sounds" or "breath sounds," are controlled by the above muscles. However, speech sounds in the very high frequency range, known as "falsettos" or "head sounds," are controlled solely by the force of exhalation.

Prior to the final emission of the sounds from the mouth, the glottal sounds pass through the pharynx, the oral cavity and the nasal cavity, during which time the tone color of those sounds is altered by the acoustic resonance within those cavities called the "vocal tract". This process constitutes the third stage of speech production. The sounds produced by the vibration of the vocal folds, called "puffing sounds," are shown as sound waves in the right part of Fig. 1.26. From this figure, it can be seen that the sound waves are triangular in shape. The waveform becomes larger as the exhalation pressure increases. However, the shape of the waveform does not change much.

The fundamental glottal sound frequency increases as the tension in the vocal fold muscles increases. Although the waveforms of the glottal sounds do not change in the vocal folds, the sounds produced from the lips exhibit a variety of tones. The reason for this is that resonance frequencies can be changed by the shape of the vocal tract. For instance, the resonance frequencies can change significantly by altering the position of the tongue inside the vocal tract. The differences in these resonance components, called "formants," characterize different vowels.

1.4.2 Resonances Inside the Vocal Tract

As shown in Fig. 1.27, which resonance frequencies are inside the vocal tract can be predicted by considering the vocal tract to be like a cylindrical tube with a length (L) of 17 cm that corresponds to the distance from the vocal folds to the lips. When the maximum amplitude of the sound wave is at the end of the tube, the sound

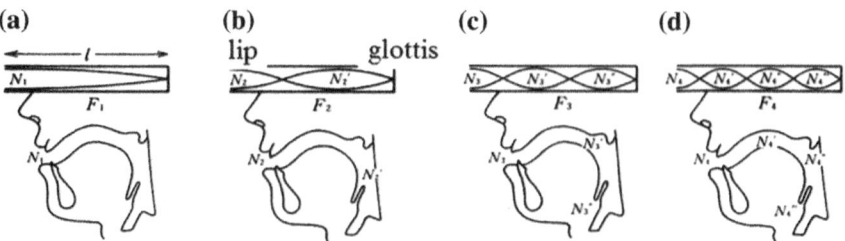

Fig. 1.27 Resonance in vocal tract

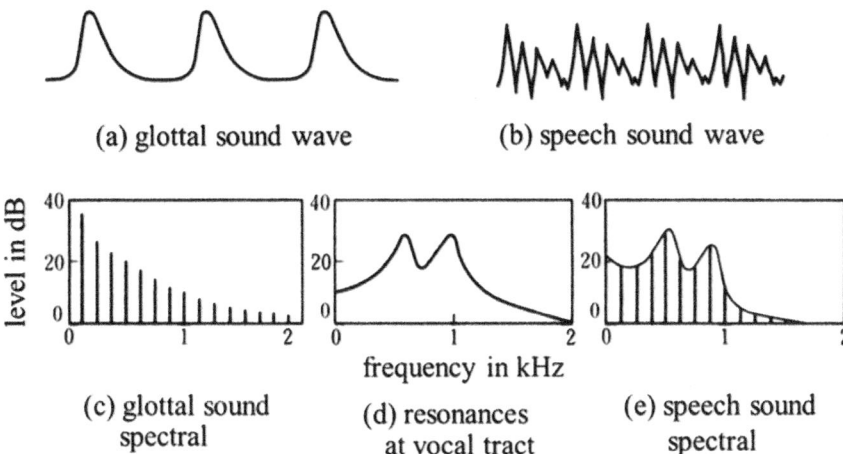

Fig. 1.28 Production process for speech power spectrum

becomes loudest. In other words, when 1/4th of the sound wavelength (λ) is equal
to the length of the tube resonance occurs. In the case of a tube with a length of
17 cm, this corresponds to a resonance frequency (f) of 500 Hz when the sound
velocity (C) is 340 m/s. It is calculated by the equation $f = C/\lambda = 34{,}000$ cm/
(4×17 cm), where $\lambda/4 = L$.

Furthermore, when 3/4ths and 5/4ths of the wavelength are equal to the tube
length, the sound intensity reaches a maximum at the end of the tube. In the same
way as above, resonances are created at frequencies of 1500 and 2500 Hz. Since the
waveform at the end of the tube is triangular, its spectrum has a harmonic structure
with multiple frequency components in addition to a fundamental frequency. Since
the waveform at the beginning of the vocal tract is also nearly rectangular in shape,
its spectral pattern decreases at 12 dB/oct, as shown in the left figure of Fig. 1.28.
Under the influence of the resonance inside the vocal tract, the corresponding
formants and their subsequent complex waveforms are created as shown in the
middle of the figure. Actually, the spectral pattern produced at the position of the
lips decreases at 6 dB/oct due to the influence of acoustic emissions.

Although the speech spectrum has an unlimited number of formants, all vowel
sounds are determined by the combination of the first and second formants. As
people can control the shape of their vocal tract, the formant frequencies can
therefore be determined at will. Even in the case of substitute speech used by people
who have undergone a laryngectomy, they can still control the formant frequencies
by changing the shape of their vocal tract. Based on this principle, various speech
substitutes have been proposed, as discussed in Chap. 6.

1.4.3 Speech Production Model

By using a cylindrical tube model, we can easily prove that the shape of the vocal tract and the waves produced through them have a one-to-one correspondence. It is known that we can estimate the shape of the vocal tract, according to the waves coming from the lips, by using a cylindrical tube model. The vocal tract has a continuous and smooth shape that, for the purpose of the model, can be divided into sections. As Fig. 1.29a shows, the tube model is made up of several disks of equal diameter that approximate the structure of the vocal tract. The left side of the figure corresponds to the glottis, while the right side corresponds to the lips. In forming this model, we usually use a total of 12 or 13 disk-shaped tubes. The parts are indicated by n, $n + 1$, $n + 2$, etc., and the corresponding areas are indicated by A_n, A_{n+1}, etc. Sound waves emitted from the left side are partially reflected at the points where the area changes. The coefficient of reflection is denoted by (1.1):

$$\alpha_n = (A_n - A_{n+1})/(A_n + A_{n+1}) \qquad (1.1)$$

$$\begin{aligned} F_{n+1} &= F_{n,p} + B_{n+1,r} = (1 - \alpha_n)F_n + \alpha_n B_{n+1} \\ B_{n+1} &= F_{n,r} + B_{n+1,p} = \alpha_n F_{n+1} + (1 - \alpha_n)B_{n+1} \end{aligned} \qquad (1.2)$$

As Fig. 1.29b shows, the components that allow sound to penetrate $F_{n,p}$ and the components to reflect sound $F_{n,r}$ are produced at points n and $n + 1$. The components that allow reflected sound to penetrate from the right side $(B_{n+1,p})$ and the components to reflect the sound $(B_{n+1,r})$ are denoted by the coefficient of reflection α_n, thereby rearranging the formula as (1.2).

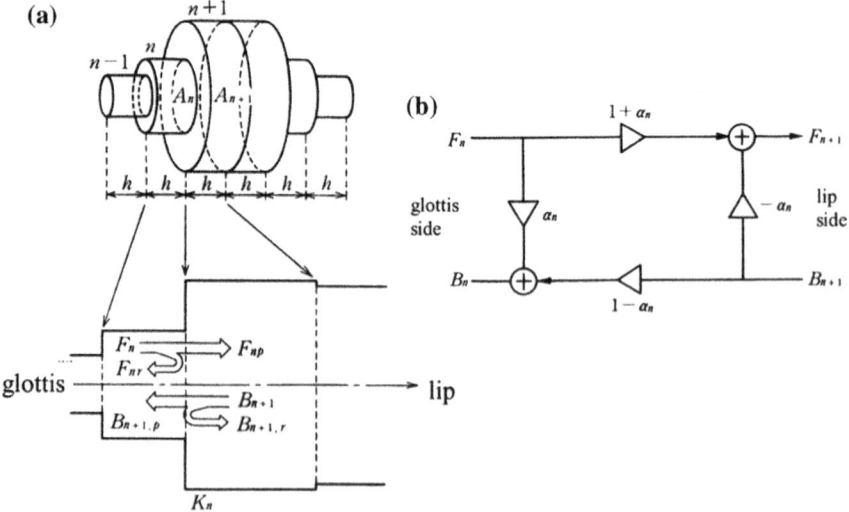

Fig. 1.29 **a** Acoustic tube model of vocal tract, **b** block diagram representation of acoustic tube model

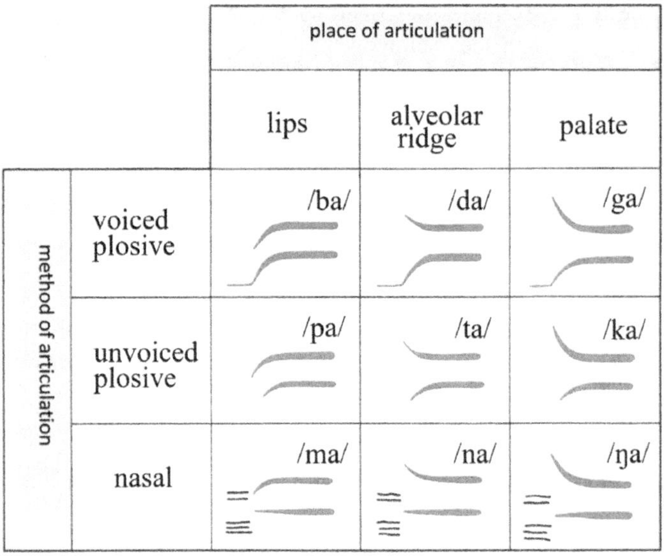

		place of articulation		
		lips	alveolar ridge	palate
method of articulation	voiced plosive	/ba/	/da/	/ga/
	unvoiced plosive	/pa/	/ta/	/ka/
	nasal	/ma/	/na/	/ŋa/

Fig. 1.30 Nine consonants classified by a place of articulation and a method of articulation. The time spectral patterns for the consonants are schematically indicated in the figure

The above formula can also be represented as shown in Fig. 1.29b.

Since we can calculate the area $(A_n, A_{n+1}, \ldots)$ of any cross section of the vocal tract by knowing α_n, it is also possible to calculate that area when we know the shape of the sound waves. In reverse fashion, by first knowing any cross-sectional area of the vocal tract, it is possible to synthesize any of the vowels. Lastly, when the vowels waveform is known, we can use that information to calculate any cross-sectional area of the vocal tract. This is the basic idea of speech analysis and synthesis methods: to make use of a few parameters to express sound information by calculating the cross-sectional area of the vocal tract from the shapes of vocal fold waves or vice versa. This principle is applied to synthesize speech sounds, especially vowel sounds, with a few parameters.

Most consonants are produced by rapidly changing the shape of the vocal tract in time. The duration from the starting point of each consonant to the following vowel is in the range from around 10 to 100 ms. Typical examples of consonants are shown in Fig. 1.30, where each consonant is classified by a "method of articulation" and a "place of articulation". The place of articulation indicates the place of "narrowing point" of the vocal tract made by lips or tongue and it also means the starting point of the movement of the lips or the tongue. The movements are strongly related to the changes of the second formant frequency as represented in the time spectral patterns of the figure. The method of articulation is largely divided into the voiced and the unvoiced consonants. In the Fig. 1.30, the unvoiced consonants are the "plosives" /p/, /t/, /k/ and the voiced consonants are both "nasal" /m/, /n/, /ŋ/ and plosives /b/, /d/, /g/.

In addition to the examples described above, the "fricatives" /sh/, the "affrica-tive" (/ts/ and /ch/), and the "semivowel" (/w/ and /y/) are used in various lan-guages. Needless to say, there are many different consonants depending on the language. For example, the "liquid sound" /l/ and the "nasal sound" /ŋ/ are Japanese-specific consonants.

PARCOR (Partial Correlation) invented by Itakura [16] is the basis of all methods in use throughout the world. Furthermore, CELP (Code Excited Linear Prediction) method developed by Schroeder and Atal is widely used for recent mobile phones [17]. Since it is possible to synthesize various sounds with a few parameters and random noises, it is also used in the production of sound synthe-sizers for speech disorders. Furthermore, it has become possible to quantitatively express speech impediments caused by abnormal shapes of the human mouth.

1.5 Maximum Transmitted Information by Human Senses

Research in sensory substitutes occupies an important position in assistive engi-neering. However, the most important factor yet to be understood is how sound can be conceptualized by means of residual senses and nervous systems. At this point, the author will not deal with the conceptual aspects of sound but rather limit his discussion to a practical understanding of the design of devices that can serve as sensory substitutes.

Generally speaking, when we substitute one sense for another, it is important to determine quantitatively how much information the substituting sense is able to transmit. To illustrate this point, we will consider the sense of touch as an example. In particular, by applying communication theory and psychological experiments, the author will describe the method used to acquire the maximum quantity of information through vibratory stimulation to two points within a certain limited area of the tactile sense. It goes without saying that the concept explained below is likewise applicable to both visual and auditory senses.

1.5.1 Category Decision Test and Channel Capacity

As Miller has indicated, the maximum transmitted information (R_t) accrued through one point of stimulation can be evaluated by means of the following equation based on a category decision test [18].

$$I(X, Y) = H(X) - H_y(X) \tag{1.3}$$

$$R_t = \max\{I(X, Y)\}$$

$H(X)$ represents the entropy of the source signal, while $H_y(X)$ equals the dissipated information. Each can be found with the following equation.

$$H(X) = -\Sigma P(x_i) \log_2(1/P(x_i)) \quad i = 1, n$$

$$H_y(X) = -\Sigma\Sigma P(x_i, y_j) \log_2\left(1/P(x_i/y_j)\right) \quad i = 1, n, j = 1, n \qquad (1.4)$$

$P(x_i)$ equals the occurrence probability of an intensity level (x_i). Here, the level (x_i) of stimulation intensity is divided into m degrees. Therefore, if all occurrence probabilities are same, $P(x_i)$ would equal $1/m$. The intensity level $P(x_i/y_i)$ represents the probability of a given intensity (x_i) when a subject replies that he/she received a stimulation of intensity (y_i). When we find the value of $I(X, Y)$ by varying the degree m, the largest $I(X, Y)$ should be designated as R_t. A conceptual diagram for the transmitted information is shown in Fig. 1.31. The author would like to explain how to determine R_t based on his experimental results [19].

In the experiment on the designation of R_t, the vibratory frequency (200 Hz)—having a sensation level of 14 dB (0 dB equals the threshold of sensation)—was equally divided logarithmically to m degree. The vibrations were applied 20 times at random to the fingertip of the normal subjects without any disabilities, at which time the subjects were asked to identify which degree of the stimulation intensity was perceived. The results of relationship between $I(X, Y)$ and m are shown in Fig. 1.32. The horizontal axis indicates m, and the vertical axis indicates the transmitted information $I(X, Y)$. It can be seen that $I(X, Y)$ reaches a saturation point when m is 5 or greater. Consequently, the maximum transmitted information (R_t) is around 1.75 bits.

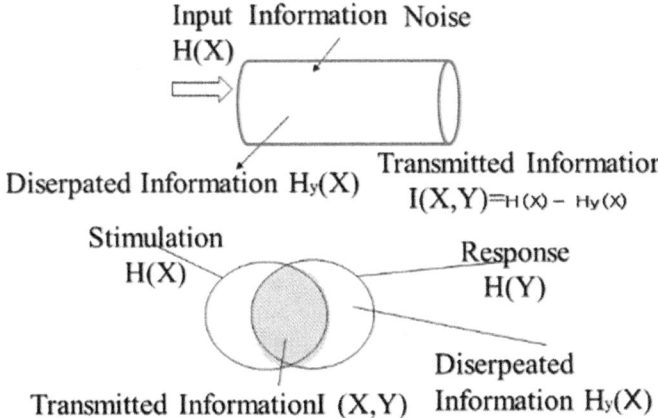

Fig. 1.31 Conceptual figure of transmitted information

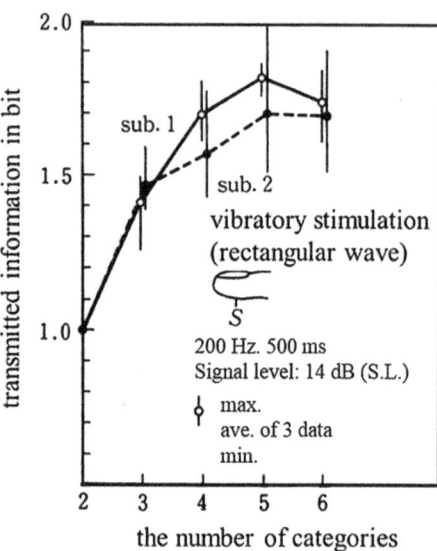

Fig. 1.32 Transmitted information $I(X, Y)$ as function of the number of categories

1.5.2 Maximum Transmitted Information for Multiple Stimulation

In this manner, R_t can be ascertained rather easily in the case of a one-point stimulation. However, when a multiple stimulation is applied simultaneously on certain limited tactile surfaces, how does R_t change? To examine this, it is necessary to measure R_t when two points are stimulated according to a category decision test.

For the experiment, two vibrations were chosen from a vibrator array arranged in 16 rows, each 1 mm apart, with 3 mm between each vibrator. Next, with one of the two vibratory stimulations, a masker stimulation of the constant intensity level was generated while the signal vibratory stimulation was overlapped with the other vibrator. In the category decision test, the intensity level of the signal was only varied at a random rate. Moreover, the intensity level of the signal was equally divided into 5 degree. Both signals and maskers were applied every 2–3 s on the fingertip of the subjects' index fingers, and the subjects were asked to identify to which of the 5 degrees the signal belonged.

As shown in Fig. 1.33, it was found that the maximum transmitted information corresponding to R_t varied greatly depending on the distance x [mm] between two stimulation points and the intensity level M [dB] of the masker. When we consider the relation between the masker level (M) and R_t, by fixing the distance x at 3 mm, it was found that R_t tends to decrease as M increases. In the case of two simultaneous stimulations, there is mechanical interference combined with the inhibition of nervous systems. This likely results in a weaker sense perception of the signal compared to the case of a single point of stimulation. Accordingly, a decrease in subjective intensity level of the signals constitutes one of the factors that reduce R_t.

Fig. 1.33 Transmitted information as function of masker level

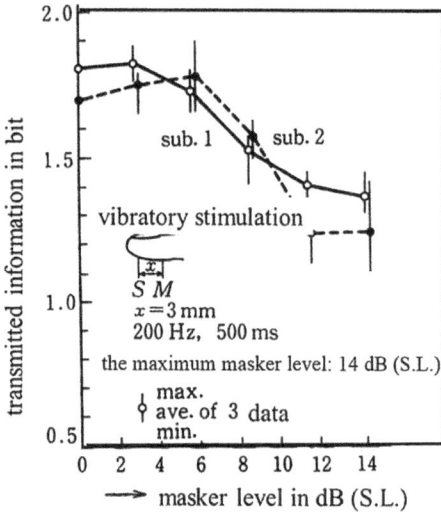

Fig. 1.34 Maximum transmitted information as function of distance between signal and masker

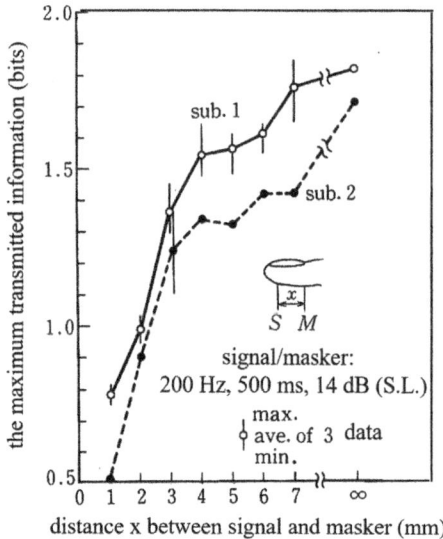

Next, as seen in Fig. 1.34, when we fixed the masker level (M) at 14 dB and determined the value of R_t by varying the distance x between the masker and the signal, it was found that R_t decreases in proportion to x. The degree of the decrease became particularly obvious when x is less than 3 mm. As stated previously, the cause of this decrease would also appear to result from the decrease of subjective intensity of the signals by the masker. Furthermore, when the distance between the

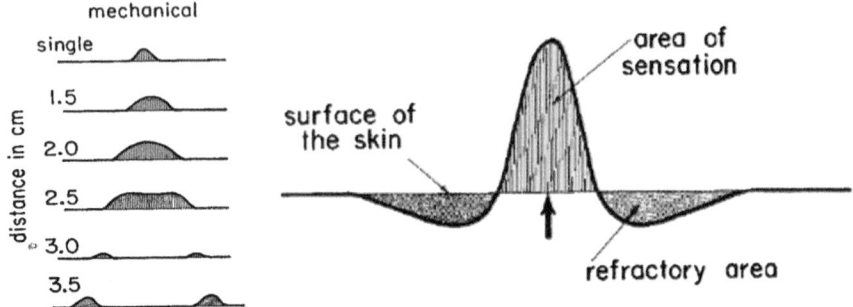

Fig. 1.35 Local distribution of sensory magnitude for mechanical stimulation with two points and its surrounding area of inhibition

two points of stimulation decreased, the overlap of the receptive field and mechanical interference became obvious. Therefore, the difficulty in perceiving the signals separately is also probably one of the causes of the decrease in R_t. The phenomenon is something that cannot be evaluated when the maximum transmitted information is derived from one point of stimulation.

The reason the maximum transmitted information decreased for two point stimulations may be explained by using the "neural unit" proposed by Békésy, as shown in the left side of Fig. 1.35. The neural unit shows the local distribution or receptive field of sensory magnitude for mechanical stimulation with two points and its surrounding area of inhibition. It is hypothesized that when two points of stimulation are added simultaneously to the tactile surface, the subjective intensity (S') and the subjective number (n') of the multiple stimulation greatly change due to the suppressive effect and the overlap of the receptive field as shown in the right side of Fig. 1.35.

This means that the maximum transmitted information may be obtained by using the estimated method mentioned above if S' and n' are determined. Therefore, if the inhibitory and receptive fields of the neural unit can be obtained by psychophysical experiments, the subjective intensity S' and the subjective number n' for the stimulation can also be estimated. One of approaches to determine S' and n' using the masking effect and the two-point threshold will be proposed in Chap. 4.

1.6 Recognition Time Difference Among the Tactile, Auditory and Visual Senses

Being different from letters or figures, sounds constitute information that changes moment by moment. As such, the question is whether the other senses can receive those sounds in real time. Stimulations such as light, sound, and vibrations require a certain time for processing in order for this information to be perceived by each sense organ via their respective sensory channels. It is preferable that the perceiving time for both the auditory sense and substitute sense be simultaneous when vocal sounds are fed back through the visual or tactile senses. It is known that even people with normal hearing may begin to stutter when they are made to hear their own vocal sounds after a delay of roughly 0.2 s. With this in mind, it is easy to imagine that the delay in perception time is a crucial factor in the control of pronunciation. However, as we have not yet found an adequate method to directly measure this basic perception time, this factor constitutes the first challenge in designing a sensory aid.

Now we will refer to the results of measuring the differences in perception time among the visual, auditory, and tactile senses. This was accomplished by presenting two stimulations at different times. Specifically, when there was a slight time gap in the pulse-like presentation of light and sounds, or when there was a slight time difference in the presentation of sounds and vibrations to an index finger, those differing stimuli were perceived to have occurred simultaneously. Furthermore, the author would like to consider whether the visual sense or the tactile sense is more advantageous as a substitute sense of the auditory sense. In order to answer this question, the author and his colleague examined both the perception time difference and the reaction time—that is to say, the speed of the vocalization reflex—between a subject's receiving stimulations and vocalizing them.

For the measurement of the perception time differences, as our source of light stimulation, we used alternating colors whose brightness was constant and which continuously changed back and forth from green to red, as shown in Fig. 1.36. We

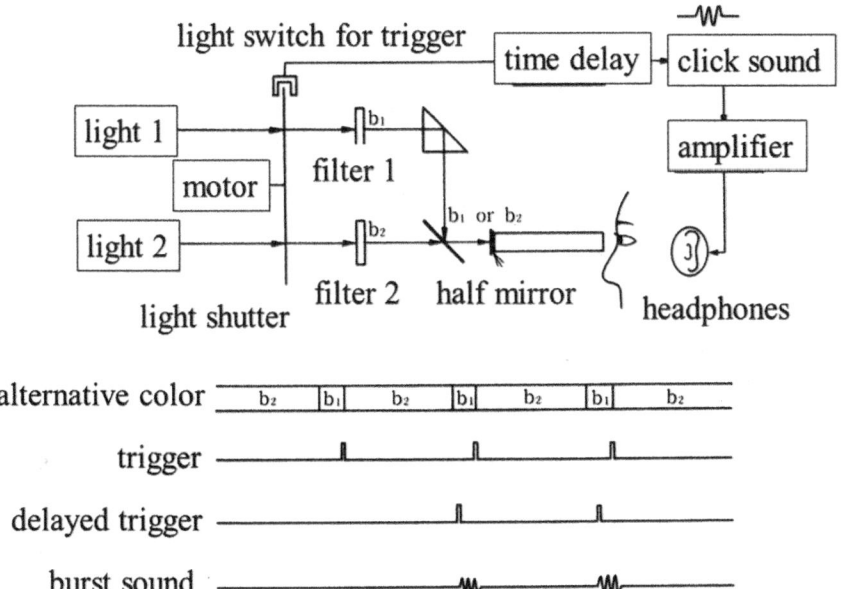

Fig. 1.36 Experimental system for measurement of recognition time difference between vision and hearing

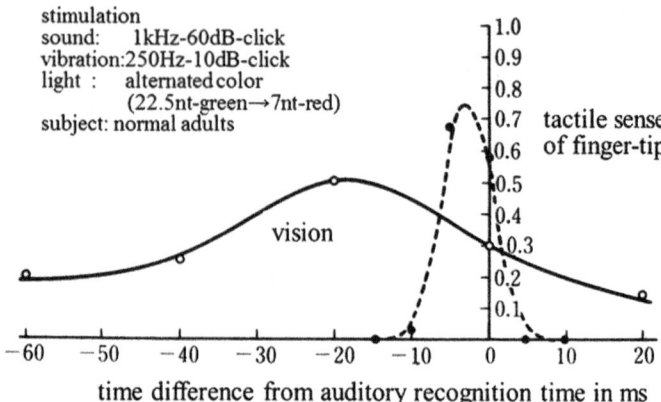

stimulation
sound: 1kHz-60dB-click
vibration:250Hz-10dB-click
light :   alternated color
            (22.5nt-green→7nt-red)
subject: normal adults

tactile sense
of finger-tip

vision

time difference from auditory recognition time in ms

Fig. 1.37 Response probability that two stimulation presented to two different senses were simultaneously perceived. The data were obtained by constant method

set the starting time of the color change as the reference point. We used single sound waves [1 kHz, 60 dB (S.L.)] for our sound stimulation, and single-wave stimulation [250 Hz, 10 dB (S.L.)] for our vibration. The measurements were conducted using a "constant method" and the subjects included people with normal hearing from 10–30 years of age.

In Fig. 1.37, the vertical axis indicates to what degree the subjects perceived the sound and vibratory stimulations simultaneously, and the horizontal axis indicates the time difference (Δt) when subjects simultaneously perceived the two forms of stimulation. The solid line represents the auditory and visual senses, while the dotted line indicates the auditory and tactile senses. From the figure, it is clear that the time delay Δt from the auditory sense to the visual sense was approximately 30 ms, while the Δt from the auditory sense to the tactile sense was also about 5 ms [20]. This fact tells us that compared to the tactile sense, the visual sense experiences a delay from the auditory sense in perception time.

The difference in the time required for perception depends mainly on the number of synapses that a stimulation goes through from the time it is received until it is perceived. It is assumed that the number of synapses to perceive visual stimulation is larger for the visual sense than for the auditory sense. Furthermore, the perception times differed according to the intensity of the stimulation. As the intensity approached the threshold, the delay in perception time reached 100 ms in the case of the visual sense while there was hardly any delay for the tactile sense.

Moreover, it is assumed that these time perception characteristics also affect the reaction time, which is to say, the period of time between the presentation of a stimulus and the receiver's vocalization. This reaction time corresponds to a time delay in the vocalization system, which is a basic factor that influences our evaluation of the combined intensity of the substitute sense and the vocal system. For the vocalization experiment, flashlight or pulse-like vibrations were presented to subjects with a random stimulation intensity. The subjects were instructed to

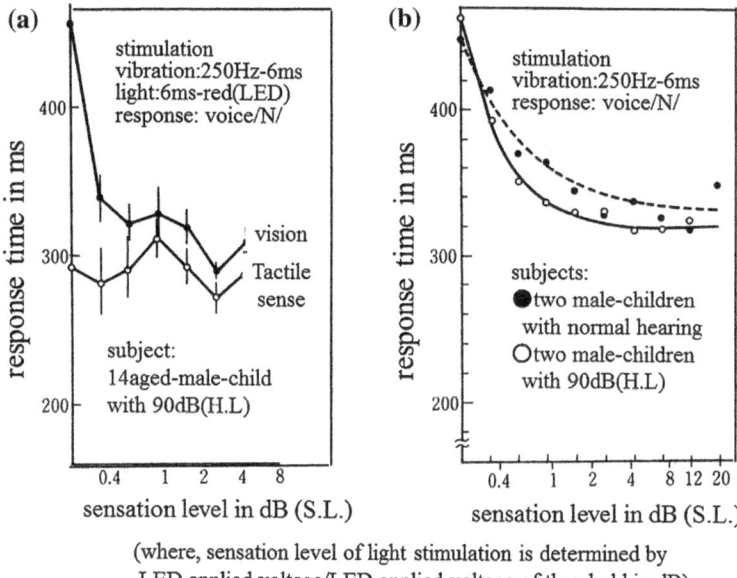

Fig. 1.38 Response time of voice /N/ as function of stimulation level. **a** Comparison of visual stimulation with tactile stimulation, **b** comparison of deaf children with normal hearing

pronounce /N/ as quickly as possible. Deaf students between 12–14 years of age participated in addition to students with normal hearing.

Figure 1.38 shows examples of experimental data of the relation between the time of the earliest sound vocalization(/N/) appearing after the sensory stimulation is introduced (vertical axis) and the stimulation's intensity (horizontal axis). Figure 1.38a is an example of a 14-year-old boy whose hearing level is around 90 dB. Figure 1.38b shows an example of two deaf boys, aged 12 and 13, and two boys with normal hearing of almost the same age. In this experiment, we measured the vocal reaction time resulting from only vibrating stimulation.

Figure 1.38a indicates that the vocal reaction time for visual stimulation is a few dozen milliseconds slower than that for tactile stimulation. In the case of deaf children, it was expected that the combined functions of the visual sense and vocalization would be strong because they have developed habits in their conversations that rely on the visual sense. This reaction time difference is an exact reflection of the perceptive time difference. As Fig. 1.38b shows, in the case of the tactile sense we did not recognize any difference between the children with normal hearing and deaf children of the same age.

Thus, from the point of view of perception time and vocalization reaction time, the tactile sense is more advantageous than the visual sense as a substitute sense for vocal sound perception and vocalization control. It is quite probable that a method could be found that would compensate for this disadvantage in perception time as the visual sense covers a much larger area of perceptive ability compared to the tactile sense. Naturally, this factor should be taken into consideration in the design of sensory aids.

In the case of auditory substitutes, as mentioned before, the important question to consider is whether it is possible to form a concept of language with voice sound information that is conveyed through other senses, and whether the hearing sense can be combined with the language concepts one has already acquired. Taking the ideas of auditory substitutes into account, from Chaps. 3–5 the author will show how speech information passes through the inside of the human brain from a point of view of studies on "aphasia," a concept known as a "speech chain" and recent progress in brain research using advanced biomedical measurements.

1.7 Language Area and the Speech Chain in the Human Brain

In 1861, Broca (Pierre Paul Broca, 1824–1880) described a patient who could understand language but could not speak. He had no motor deficits to account for his inability to speak. He could whistle, utter isolated words, and sing the lyrics of a melody. Postmortem examination of his brain showed a lesion in the posterior region of the frontal lobe. This region is now called "Broca's area." A few years later, in 1876, Wernicke (Karl Wernicke (1848–1905) described another type of aphasia. This aphasia, or language disorder, involved a failure to comprehend language rather than a failure to speak. The location of the lesion in Wernicke's patient was different from that in Broca's patient. It was at the junction of the temporal, parietal and occipital lobes-now called "Wernicke's area." The junction of temporal, parietal, and occipital lobes is an important area to remember [21].

Wernicke proposed that language involves separate motor and sensory programs located in different cortical regions. The motor program, located in Broca's area was suitably situated in front of the motor area that controls the mouth, tongue, and vocal cords. The sensory program, located in Wernicke's area, was suitably surrounded by the posterior association cortex that integrates auditory, visual, and somatic sensations.

Wernicke's model is still referred to today. According to this model, initial processing of spoken or written words takes place in the primary and unimodal sensory areas for vision and audition. This information is then conveyed to the "angular gyrus" of the posterior association area. This was thought to be the area where either written or spoken words were transformed into a common neural representation. Then, these representations were thought to be transferred to Wernicke's area where they were recognized as language and associated with meaning. Without meaning, there can be no language comprehension. These neural representations along with their associated meanings are then passed along, via the "arcuate fasciculus", to Broca's area where it is transformed into a motor representation that allows for speech.

As Denes and Pinson describe in their book "Speech Chain," speech communication consists of a chain of events linking the speaker's brain with the listener's brain [22]. They call this chain of events the "*speech chain*." As shown in Fig. 1.39, the transmission of a message begins with the selection and ordering of suitable words and sentences. This can be called the "*linguistic level*" of the speech chain. The speech event continues on the "*physiological level*", with neural and muscular activity, and ends on the speaker's side with the generation and transmission of sound waves, the "*physical (acoustic) level*" of the speech chain.

At the listener's end of the chain, the process is reversed. Events start on the physical level, when the incoming sound waves activate the hearing mechanism. They continue on the physiological level with neural activity in the hearing and perceptual mechanisms. The speech chain is completed on the linguistic level when the listener recognizes the words and sentences transmitted by the speaker. The speech chain, therefore, involves activity on at least three levels—linguistic, physiological and physical—first on the speaker's side and then at the listener's end.

Due to recent progress in non-invasive measurement technologies such as functional MRI and positron emission tomography (PET) of human cerebral cortex, the mechanisms operating in the human auditory cortex have been gradually made clear. The locations of Broca's area and Wernicke's area on the cerebrum have been identified and also the speech chain model has been established although it was found that there are other brain areas responsible for different aspects of language (Fig. 1.40). The technologies and findings should be utilized to design assistive devices for people with diseases of the language cortex and the auditory cortex including the sensorineural hearing impaired and aphasia as described in Chap. 6.

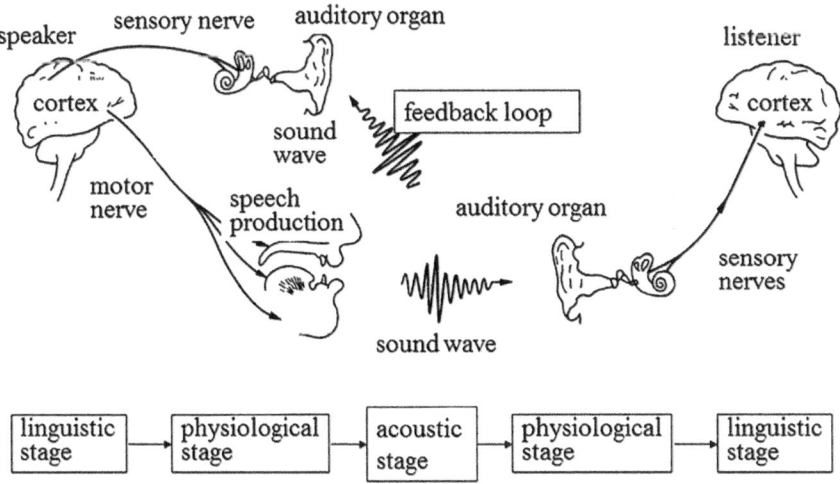

Fig. 1.39 Speech chain

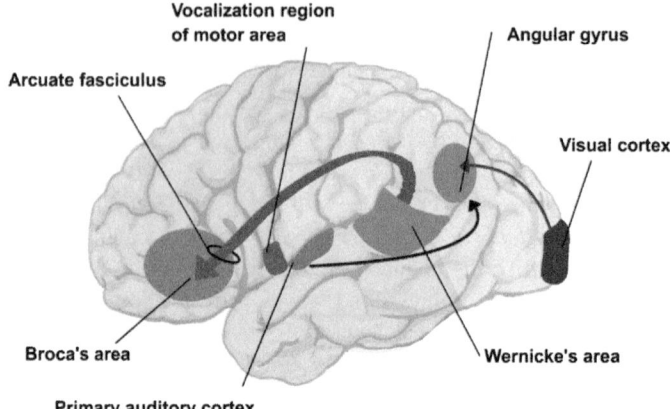

Fig. 1.40 Speech area based on cortex excision

References

1. T. Ifukube, Sound-based assistive technology supporting "seeing", "hearing" and "speaking" for the disabled and the elderly. Key note speech, in *Proceedings of the INTERSPEECH2010:11–19* (2010)
2. T. Ifukube, A neuroscience-based design of intelligent tools for the elderly and disabled, in *Proceedings of the 2001 EC/NSF Workshop on Universal Accessibility of Ubiquitous Computing: Providing for the Elderly (WUAUC'01)* (ACM Press, 2001), pp. 31–36
3. T. Ifukube, *Challenge of Assistive Technology* (Chuokoron-Shinsha, 2004), p. 12 [in Japanese]
4. N. Wiener, *Cybernetics, or Control and Communication in the Animal and the Machine* (Princeton University Press, 1948)
5. *Annual Report on the Aged Society* (Cabinet Office, 2012)
6. https://www.jst.go.jp/s-innova/research/h22theme05.html (2016)
7. R. Hayashi, J. Isogai, P. Raksincharoensak, M. Nagai, Autonomous collision avoidance system by combined control of steering and braking using geometrically-optimized vehicular trajectory. Veh. Syst. Dyn. **50**(Suppl), 151–168 (2012)
8. R. Raksincharoensak, Katsumi, M. Nagai, Reconstruction of pedestrian/cyclist crash-relevant scenario and assessment of collision avoidance system using driving simulator, in *Proceedings of 11th International Symposium on Advanced Vehicle Control (AVEC)* (2012)
9. M. Hirose, Koreisha cloud no kenkyukaihatsu (Research and development of senior cloud). Trans. Virtual Reality Soc. Jpn. **19**(3), 21–25 (2014) [in Japanese]
10. A. Hiyama, M. Kobayashi, H. Takagi, M. Hirose, Collaborative ways for older adults to use their expertise through information technologies. ACM SIGACCESS Newslett. **110**, 26–33 (2014)
11. Article in Newspaper (1901) 20th century prophecies. Hochi Shimbun (Houti Newspaper) [in Japanese]
12. E.M. Wenzel, Three-dimensional virtual acoustic displays, in *Multimedia Interface Design*, ed. by M.M. Blattner, R. B. Dannenberg (ACM Press, 1992), p. 257
13. G. Békésy, Neural funneling along the skin and between the inner and outer hair cells of the cochlea. J. Acoust. Soc. Am. **31**(9), 1236–1249 (1959). doi:10.1121/1.1907851
14. N. Suga, Sharpening of frequency tuning by inhibition in the central auditory system: tribute to Yasuji Katsuki's paper. Neurosci. Res. **21**(4), 287–299 (1995)

15. P. Gilad, S. Shtrikman, P. Hillman, M. Rubinstein, A. Eviatar, Application of the Mössbauer method to ear vibrations. J. Acoust. Soc. Am. **41**(5), 1232–1236 (1967)
16. F. Itakura, S. Saito, On the optimum quantization of feature parameters in the PARCOR speech synthesizer, in *Proceeding of the Conference on Speech Communication and Processing*, pp. 434–437 (1972)
17. M. Schroeder, B. Atal, Code-excited linear prediction(CELP): high-quality speech at very low bit rates. Acoust. Speech Signal Process. IEEE Inter. Conf. ICASSP '85. **10**, 937–940 (1985). doi:10.1109/ICASSP.1985.1168147
18. G.A. Miller, The magical number seven, plus or minus two: some limits on our capacity for processing information. Psychol. Rev. **63**(2), 81 (1956)
19. T. Ifukube, Maximum information transmission on a tactual vocoder: In the case of time invariant stimulation. Jpn. J. Med. Electron. Biol. Eng. **17**(3), 230–236 (1979). doi:10.11239/jsmbe1963.17.230. [in Japanese]
20. T. Ifukube, Artificial reality based on biomedical engineering—as an example case of sensory substitute studies. The J. Inst. Telev. Eng. Jpn. **46**(6), 718–726 (1992). doi:10.3169/itej1978.46.718. [in Japanese]
21. P. Wilder, R. Lamar, *Speech Chain and Brain Mechanism* (Princeton Legacy Library, 1965)
22. P.B. Denes, E.N. Pinson, *The Chapter 4 Speech Chain: The Physics and Biology of Spoken Language* (1973)

Chapter 2
Sound Signal Processing for Auditory Aids

Abstract In this chapter, signal-processing methods for digital hearing aids are discussed and the author also describes how auditory characteristics and speech understanding abilities change because of hearing impairment as well as aging. In particular, the author introduces various approaches to signal processing such as noise-reduction methods, consonant-enhancement methods, and speech rate or intonation-conversion methods. Past and recent artificial middle ears are also discussed as one form of hearing aid.

2.1 Kinds and Characteristics of Hearing Impairment

Hearing impairment presents various symptoms depending on such factors as the defective area, the degree of impairment, and the general characteristics of such impairment. Specifically, the hearing impaired can be categorized into two main groups: the "organically impaired", and those who are "functionally impaired", as shown in Table 2.1. The former group is further subdivided into individuals with "conductive hearing impairment"—characterized by a defect in the ability to convey sounds through mechanical signals from the middle ear—and the "sensory neural hearing impaired", whose disability results from a deeper part of the inner ear which includes hair cell or nerve channel malfunctions.

Among the sensory neural hearing impaired, inner ear auditory disorders are characterized by impairment that is limited to the cochlea, while the impairment experienced by individuals with "retro-cochlear" auditory disorders consists mainly of malfunctions of the cochlear nerve or, in certain cases, the central nerves. The terms "central nerve deafness" and "cortical deafness" are also used to indicate malfunctions of the central channel of the auditory sense and the cerebral cortex, respectively. However, the impairment within these areas differs from the ordinary concept of defective hearing and should, more specifically, refer to people who have an "abnormal auditory sense".

The results of hearing tests are shown in Fig. 2.1 as an "audiogram", which has a fixed form. Presently used audiograms indicate the frequency in logarithmic scales

© Springer International Publishing AG 2017   43
T. Ifukube, *Sound-Based Assistive Technology*,
DOI 10.1007/978-3-319-47997-2_2

Table 2.1 Classification of hearing impairments

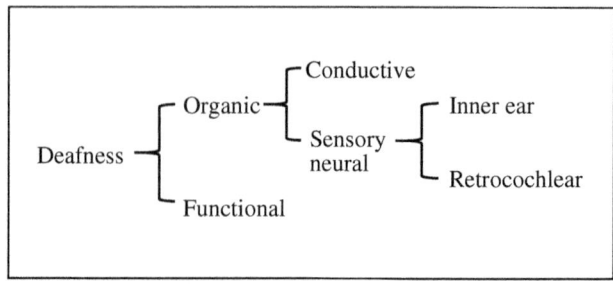

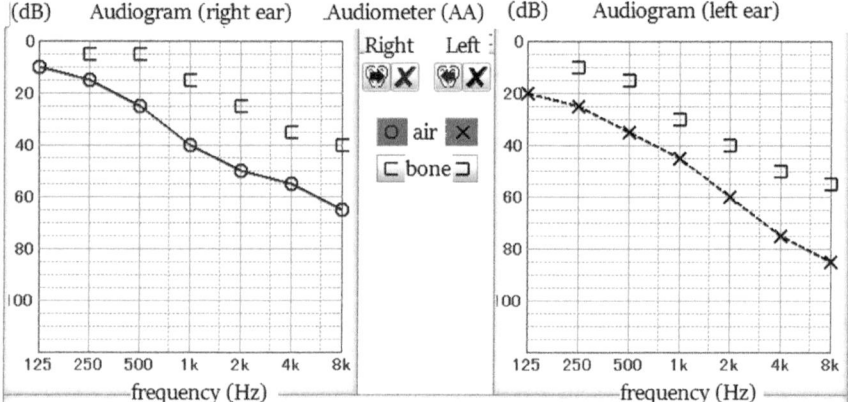

Fig. 2.1 An example of hearing level (vertical axis in dB) versus sound frequency (Hz)

on the horizontal axis, and the hearing level in dB as scales on the vertical axis. The "air-conduction hearing level" is a threshold level measured when sound is sent directly through headphones, and the "air-conduction audiogram curve" shows a curve made by connecting those dots. It reflects the conductive and/or sensory neural hearing impairments. Whereas, "bone-conductive hearing level" is a threshold level when vibratory stimulation is given to a papillary portion (bone on the back of the pinna) through a bone conduction receiver, and it reflects sensory neural hearing impairments.

It goes without saying that the further this curve decreases from the 0 dB line, the more defective hearing becomes. The limit of normal hearing is generally considered to be 10 dB (JIS standard value). However, normality and abnormality should not be exclusively judged from recorded dB levels; age, test situation and subjective symptoms of subjects should be also considered with some degree of flexibility. In order to grasp the general tendency of air-conduction hearing ability and to connect it with social adaptation ability, sometimes the hearing ability of a

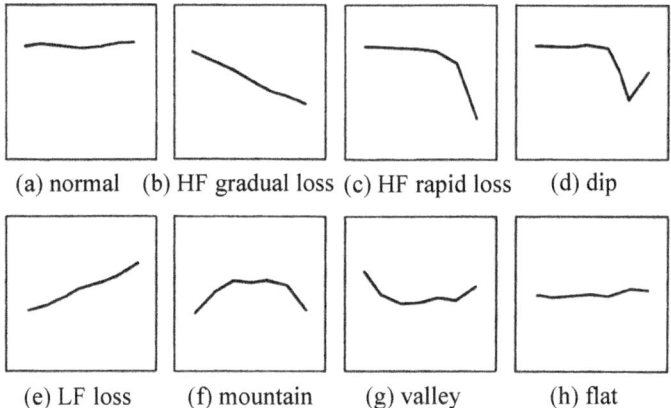

Fig. 2.2 Eight types of hearing loss by frequency

subject's ears is indicated by one dB level. This is called the "average hearing level." In Japan, the commonly used method is to fix each measured level of 500, 1000, and 2000 Hz as a, b, and c, respectively, with the average hearing ability indicated as (a + 2b + c)/4 [dB].

The curve of the air-conduction audiogram varies, as seen in Fig. 2.2. Moreover, when these curves are arranged in order, they are called "hearing types." This approach is not only convenient for briefly expressing the hearing ability of individual cases, but also to diagnose diseases when such diseases are associated with a specific type of hearing ability.

Defective hearing is generally thought to be merely a weakened hearing ability that can be recovered through the use of hearing aids, in the same way as wearing glasses improves eyesight. Specifically, glasses compensate for optical deformation; however, in the case of the hearing impaired, a hearing aid corresponds to a widening of the auricle to merely increase sound volume. However, many hearing-impaired people experience not only sound system deterioration, but also suffer from disorders in unknown parts, such as the inner ear or the central nerves. As a result, hearing aids whose function is limited to the amplification of sound pressure often lead to unsatisfactory results.

2.2 Auditory Characteristics of the Hearing Impaired

2.2.1 Recruitment

Recruitment is a phenomenon of disease symptoms that audiograms cannot detect and which can occur among certain hearing-impaired individuals who experience a considerable degree of hearing loss. While those people are hearing sounds above

Fig. 2.3 Description method
of recruitment

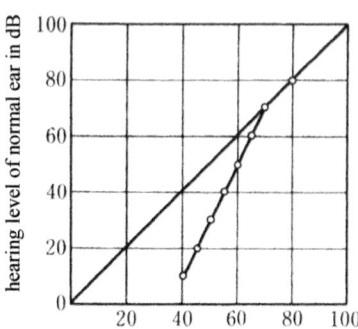

threshold, their sense of sound intensity increases at an abnormally faster rate than
for people with normal hearing when the intensity of those sounds is gradually
increased, until finally hearing-impaired individuals are able to hear the sounds at
the same volume as people with normal hearing. Figure 2.3 shows this phenomenon
by indicating the actual sound intensity on the horizontal axis and the subjective
sound intensity on the vertical axis. In the case of people whose ears suffer from
recruitment, the range between their actual threshold of hearing and their uncomfortable threshold is clearly narrower than for people with normal ears.

Therefore, when weak sounds and strong sounds are mixed in a complicated way
—as in the case of the elements of spoken words—ears suffering from recruitment
cannot detect the sounds as accurately as normal ears. Thus, the total discrimination
score does not significantly improve. This is because the weak components of
sound are still inaudible to the listener despite their amplification through hearing
aids, whereas strong sound components enter the ears with excessive intensity [1].
Compensating for this recruitment factor also constitutes an important theme in the
development of digital hearing aids—a theme that will be explained in greater detail
later. For patients with sensory neural hearing impairment—characterized by
impairment of the inner ears or the central nervous system—there is a possibility
that their auditory sense properties necessary for the extraction of voice sound
characteristics have also deteriorated. Therefore, it follows that the hearing of voice
sounds does not necessarily improve by merely amplifying sound intensity.

2.2.2 Combination Tones

Combination tones are a phenomena that occurs when sounds with a frequency of f_1
and f_2 are heard simultaneously, thus creating other sounds $(2 \times f_1 - f_2, f_2 - f_1)$,
which are heard mostly in the inner ear. These new sounds may be termed "imaginary
sounds" as they do not actual exist in the physically sense but rather are generated in the
inner ear, as shown in Fig. 2.4. In particular, the imaginary sound $(f_2 - f_1)$ is called a
"difference sound." As the process of electrical conversion by the hair cells is also

Fig. 2.4 Schematic figure
showing production of
combination tones

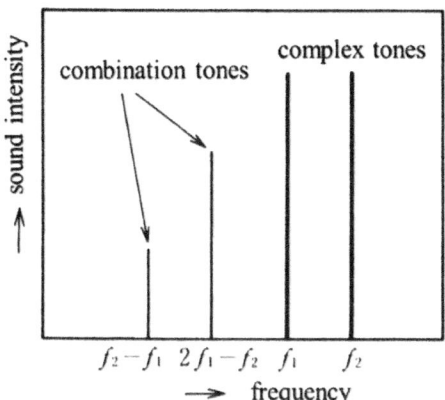

nonlinear, that too is related to the generation of combination tones. "Otoacoustic emissions" are a process assumed to be created in the Corti organ inside the cochlear where the ear emits single sounds. Also, when people observe the output of single-sound otoacoustic emissions, the level of combination tones increases.

As will be described in Chap. 6, although there are considerable peaks that correspond to the second and third formants in the spectra of Mynah birds that can imitate human voices, most of them have either no first formants at all or only extremely faint ones. However, as human beings can distinguish vowel sounds by the difference in combinations of the first and second formant frequencies—even in the case of a mynah's sounds—there is a possibility that the components of the first formant can be generated in the human auditory sense system. In considering how the first formant is produced in the auditory sense system, one assumption is that this perception results also from combination tones created by the existence of the second and third formants appearing near the first formant.

If these combination tones are in fact essential to the hearing of voice sounds, one possible idea would be to develop new hearing aids that can artificially add such combination tones by means of signal processing. If successful, this would clearly benefit hearing-impaired people who cannot generate combination tones on their own.

At this point, the author would like to offer a description of the nonlinearity of the auditory sense and its effects on hearing. In fact, one difference in perception between people with normal hearing and the hearing impaired is with respect to this nonlinearity. Specifically, when the hearing impaired are exposed to combined sounds that undergo excessive change, such as voice sounds, what they actually hear is different from those with normal hearing. This fact is an important consideration in determining which kinds of algorithms are most effective in the design of digital hearing aids.

Although the two mainstream methods for the measurement of combination tones are the cancellation method and the otoacoustic emission method, these methods are ineffective when attempting to examine combination tones while a person is hearing complex tones whose frequency changes like voice sounds.

Therefore, by using the fact that the masking level in FM tone increases, Imamura and his colleagues examined the perceptual characteristics of combination tones with a masking method that utilizes complex FM tones as a masker and short sounds as probe tones. More concretely, by using the masking approach, they were able to indirectly estimate the level of combination tones [2].

Although the details are omitted, there is a possibility that combination tones, especially difference tones, are more clearly perceived by people with normal hearing for sounds whose harmonic elements continuously undergo frequency modulation (FM), such as in the case of the formant transition of consonant voice sounds. However, in the case of the hearing impaired, there was no perception of difference for both regular complex tones and FM complex tones. These results clearly tell us that, unlike those with normal hearing, the hearing sense of the hearing impaired is characterized by a different kind of nonlinearity.

Although the role that combination tones play in voice sound perception remains unknown, it can be said that, compared to people with normal hearing, the hearing impaired do indeed perceive voice sounds differently as it is difficult for the hearing impaired to generate combination tones.

2.2.3 Speech Articulation Curve

Needless to say, the largest role played by the hearing organs in daily life is to enable people to hear what others are saying. Therefore, in the auditory sense field, the testing of speech hearing ability has been emphasized as a means of judging social adaptation ability. Since the precise measurement of hearing ability became possible with the advent of speech audiometry, this method of evaluation has been considered the most effective for the assessment of not only conversational hearing ability but also for diagnosing which parts of the auditory system are defective. The most standardized methods clinically practiced today include measuring the speech perception threshold as well as calculating the speech discrimination score.

In testing speech discrimination, a pair of words are first chosen from a test chart and subjects are asked to write them down after hearing them at an easily detectable sound level. Next, the sounds are gradually weakened by 10–20 dB. The correct answer rate is then calculated for each dB level and plotted as dots, which form what is known as a "speech articulation curve" (Fig. 2.5). The highest point on the curve corresponds to the best answer rate and constitutes that person's speech discrimination score. In this way, the speech audiogram enables the tester to produce a chart that illustrates one's speech articulation curve.

Much information can be gained from analyzing such speech articulation curves. Specifically, in the case of normal ears, the speech discrimination score equals 100% or a value close to it. Figure 2.5 offers four patterns of articulation curves, with the curve on the left representing normal hearing. Compared with normal ears, pattern A is slightly to the right, but forms almost the same rising curve as in the

Fig. 2.5 Speech articulation curve

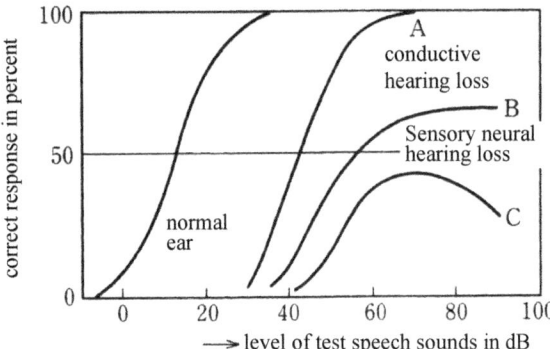

case of normal ears, with the speech discrimination score also reaching 100%. This shape is observed in the case of conductive deafness and indicates that such a person should have little difficulty carrying on a conversation if sounds are intensified through the use of hearing aids. On the contrary, the speech discrimination score in pattern B does not rise above 66% and becomes horizontal even following an increase in the intensity of the sounds. In the case of pattern C, however, rather than a rise in the articulation curve, which normally corresponds to increased sound intensity, the result indicates that the articulation curve actually falls.

The shapes of both B and C are found in cases when the sensory neural system is defective. In particular, retro-cochlear auditory disorders present a distinct decrease in discrimination ability compared to pure tone hearing loss. In the case of such defective ears, not only does the threshold raises but also there is a distortion that accompanies the perception of sounds above the threshold. In other words, the louder the sound, the greater the distortion, such that speech discrimination becomes increasingly difficult. This is a phenomenon peculiar to hearing loss among the elderly.

In this way, although the method of diagnosis has been somehow established, the greatest problem is how best to deal with central nerve deafness; characterized by defectiveness in either the central auditory system or the language cortex. In order to resolve this issue, an attempt has been made to design digital hearing aids with a built-in computer that can compensate for the defect by taking over the signal processing normally accomplished by the auditory sense. Specifically, the goal of such computer-enhanced digital hearing aids is to improve hearing by performing the central auditory system's function of controlling noise in order to extract only those important voice sounds necessary for speech comprehension.

At this point, however, suffice it to say that such digital hearing aids did not prove to be very successful, as they could not clearly outperform conventional analog hearing aids, at least for the normal hearing sense. In fact, in some cases their use actually made it harder to hear certain words. The reason for this problem is that still relatively little is known about the actual mechanism used by the human auditory sense channel—extending from the inner ear to the central nervous system—as it processes speech sounds. Consequently, it is still not easy to

Fig. 2.6 Change in average
air-conduction audiogram
with age

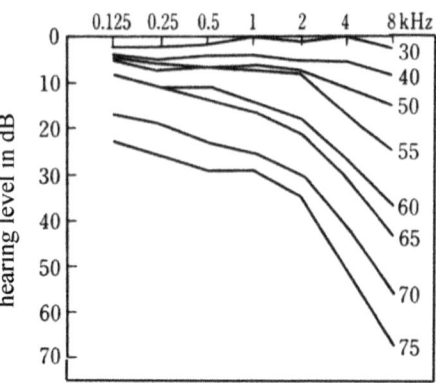

determine how to best design the software for such digital hearing aids. In the final analysis, until the question of how speech sounds are processed is clearly answered, it will be difficult for digital hearing aids to offer much benefit to users.

As shown in Fig. 2.6, the hearing level particularly in higher frequency than 2 kHz is decreasing with the age. Since the elderly disordered are recently increasing in the world as well as in Japan, judging from this increase, there is a possibility that at some point in the future, 70% of hearing-impaired people might be unable to obtain employment. In view of this alarming development, research to finally understand the mechanism of speech sound processing in order to improve digital hearing aids will become even more essential.

Although it may appear to be a roundabout route, by thoroughly analyzing the differences between the auditory sense characteristics of people with normal hearing and hearing-impaired people, we will gradually come to know how to best program software in the design of digital hearing aids.

2.3 Assistive Methods Using Hearing Aids

2.3.1 Presently Available Kinds of Hearing Aids

In the extremely rare case that a person is born completely deaf, the use of a hearing aid offers some benefit. However, even among the adult hearing impaired who have retained some residual hearing, many can expect only limited benefit through the use of hearing aids. Fortunately, most hearing-impaired children are born with residual hearing ability within a certain range. Therefore, hearing aids can play an important role for such hearing-impaired children, provided that they receive early training in hearing ability. As for the hearing impaired in general who have retained some residual hearing ability, various models of hearing aids can offer real benefit in their daily lives.

For many people with conductive deafness, it is possible to recover their auditory sense through surgical treatment. As the conductive system is almost linear, it is relatively easy for them to compensate for the linear amplification characteristics through the use of hearing aids. However, in the case of sensory neural hearing impairment, such direct treatment is almost impossible. Moreover, it is very difficult to choose hearing aids that are suitable for each individual. This is due to the possibility of complications in hearing defects such as hearing sense distortion, recruitment phenomenon, temporal resolution, and degradation in frequency resolutions. In fact, it is not rare to find hearing-impaired people who are reluctant to use their hearing aids because of the discomfort caused when their particular kind of hearing aid has been poorly chosen.

The basic components of a hearing aid consist of a microphone, a filter, components to control amplification gain and frequency response, a receiver or earphone, and a storage battery. Fortunately, progress in miniaturization technology has made these components considerably lighter and smaller than they once were. Recently, microminiaturization has made it possible to develop hearing aids such as the ITC (In-The-Canal) which, as the name implies, are actually inserted into the external ear canal.

Additionally, there is a bone-conducting type of hearing aid that conveys vibrations directly to the cochlea by having vibrators touch the mastoid bone behind the ears. Although this was developed and widely used for conductive deafness—characterized by defectiveness in the middle ear—it fell into disuse due to advances in surgery. Recent innovations have also witnessed the development of implantable hearing devices, which directly stimulate the cochlea by vibratory stimulation offer the advantage of avoiding sound feedback.

Some devices function by vibrating magnets that have been placed on the tympanic membrane or the auditory ossicles by use of an induction coil. Still other implantable hearing devices involve the insertion of a pin into the temporal region to electromagnetically convey stimulations through bones or by using vibrators. These methods are called "artificial middle ears," and will be mentioned again at the end of this chapter.

Recently, advances have also led to an increase in both the use of artificial middle ears as well as hearing aids capable of performing various nonlinear processes through the use of microcomputers. In fact, sufficient progress has been made with these devices—especially digital hearing aids—to finally make their practical use a reality. At this point, the author would like to describe the types of auditory compensation possible through the use of present-day digital hearing aids.

2.4   Digital Hearing Aids

In the case of sensory neural hearing impairment, the characteristics vary depending on the person. In particular, some such variations include a decrease in the dynamic range resulting from an increase in the hearing threshold, non-linearization of the characteristics of the input and output data as well as deterioration in both the frequency and temporal resolutions. Therefore, it is difficult to choose suitable hearing aids for each individual based merely on the simple process of amplification.

Although advances in digital technology have recently made subtle adjustments possible, sufficient knowledge as to what particular process is most effective to compensate for each individual's hearing loss has not yet fully developed. As a case in point, contrary to expectations, it is necessary to consider the possibility that emphasizing voice sound characteristics could actually prove detrimental to a user's comprehension.

However, the recent progress in digital hearing aids has offered new possibilities for the improvement of auditory sense compensation. One great advantage of digitalization is that it enables assistive devices to be adapted to the characteristics of each hearing-impaired individual by merely altering the program. In addition, voice sounds conveyed by signal processing can be emphasized more quickly and flexibly.

In short, such recent breakthroughs have led to the development of signal-processing technology that specifically targets voice sounds, while at the same time, taking advantage of the new methods for controlling noise. In so doing, increasing importance will be paid to improvements in the selection and evaluation of individual hearing aids. However, real progress in hearing assistive technology must also be accompanied by careful discussion regarding the selection method, adaptability measures and appropriate methods of instruction for users combined with the necessary counseling.

Figure 2.7 shows the general structure of digital hearing aids made by Siemens Co. Ltd. As mentioned previously, much of present-day research focuses on

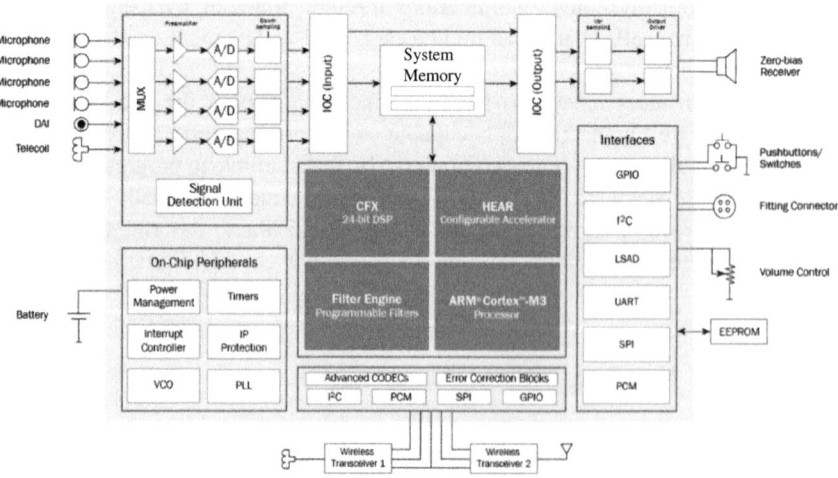

Fig. 2.7 General structure of digital hearing aid (from Siemens Co. Ltd.)

developing low-power consumption miniaturized chips. Since clear progress has already been realized in this area, it has been possible to manufacture ear canal-type digital hearing aids. It is surprising, looking at the figure, to see how many chips are used in a tiny digital hearing aid.

By making use of a microprocessor, digital hearing aids can more precisely and more flexibly adjust various parameters with the aid of an externally connected computer. Furthermore, in addition to making it possible to emphasize various sound components in order to facilitate the understanding of voice sounds, digital technology has also seen progress in helping to resolve two other common problems associated with the older analog hearing aids: namely, the ability to reduce or eliminate interference caused by different kinds of background noise and the capacity to suppress or eliminate "howling"—often described by users as an irritating "whistling sound".

Rapid progress is also being made in research to develop hearing aids that are programmed with CPUs. As described in Chap. 4, the hearing impaired often experience a phenomenon known as "increased forward masking," whereby consonants following vowel sounds are masked by those vowel sounds, thus reducing one's hearing ability. Furthermore, it is also assumed that such increased forward masking may lower frequency resolutions, thereby resulting in poor formant selectivity. Another apparent problem is that as present-day analog hearing aids increase environmental noise in addition to voice sounds when the user is in a noisy area, the hearing aid can actually increase the difficulty in distinguishing voice sounds. Present-day digital hearing aids need to compensate for these difficulties by use of digital signal processors.

The most basic type of defective hearing involves a phenomenon characterized by a narrowing in the dynamic range of the auditory sense due to a rise in the threshold of hearing. Another hearing defect is known as the recruitment phenomenon as mentioned above, which is characterized by a sudden and disproportionate increase in perceived loudness following only a slight increase in sound input. In summary, the aim of most conventional digital hearing aids has been to compensate for such characteristic defects.

2.4.1 Speech Processing for Digital Hearing Aids

2.4.1.1 Multiband Amplification Compression

In an attempt to adapt digital hearing aids to the different levels of bandwidth, research has been conducted to determine the optimal forms of "multiband amplification compression". This process involves first dividing a frequency band into several bands, followed by amplification and compression of each band, as shown in Fig. 2.8. It has been confirmed that real-time processing can be accomplished by designing an amplification compressor for multiple bands that is equipped with "mirror multiplication." Moreover, by utilizing amplification

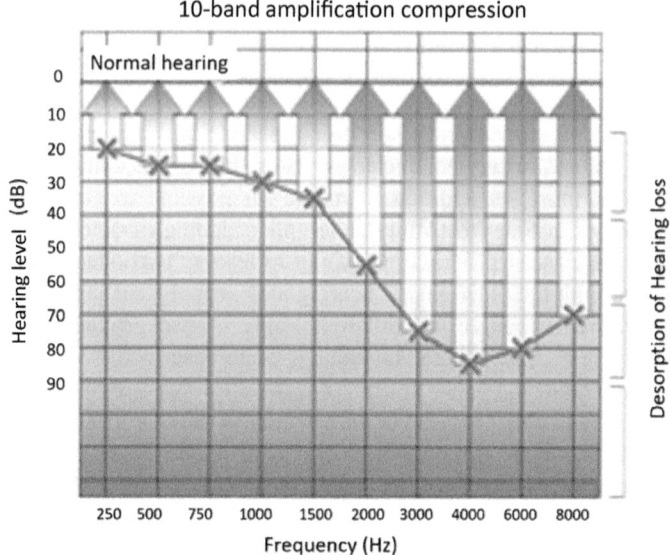

Fig. 2.8 Multiband amplification compression (10 bands in this case)

compression, gain can be adjusted as a function of frequency at each input signal level. This promising multiband amplification compression method has been adopted in recent digital hearing aids.

Furthermore, in the field of hearing compensation by means of digital hearing aids, various methods of compression have been considered to ensure that the sound pressure output signal level does not go beyond a person's comfortable volume level as a result of excessive amplification. More specifically, although ordinary amplification is applied to low sound pressure, high sound pressure amplification must be limited to ensure a comfortable sensation for the user. By using this approach, distortion is lower than in the case of simple peak clipping and there is less of a decrease in sound recognition.

2.4.1.2 Recruitment Compensation

The auditory dynamic range of the sensory neural hearing impaired is narrow in general, with the recruitment phenomenon often being observed. Therefore, as a means of compensation, a proposal was made to use CPUs to calculate the average spectrum of each octave as shown in Fig. 2.9a, b [3]. It was proved that by using a compression method with a filter whose characteristics change over time, the recruitment phenomenon could then be resolved by allowing the hearing impaired to gain the same degree of loudness as those with normal hearing. It was also found that this method resulted in less distortion than the conventional methods that used a

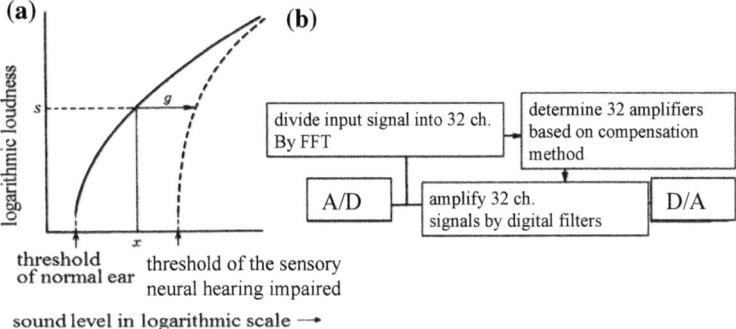

Fig. 2.9 a Compensation method of recruitment, **b** digital hearing aid with recruitment compensation function

filter bank. In view of such promising developments, this improved process has been put into practical use for most digital hearing aids.

One such method of recruitment is called "frequency shaping." The goal of this process is to compensate for a decrease in the dynamic range caused by the recruitment and to adjust that compensation to best fit the needs of each individual's characteristics. For example, one alternate approach is to set the frequency gain characteristics in such a way as to eliminate any sense of discomfort when hearing sounds uttered throughout the range of ordinary speech.

2.4.1.3   Automatic Gain Control

The aim of automatic gain control (AGC) is to adjust the gain so that all sounds will remain in the zone of a person's residual hearing ability. This process makes use of fixed values for a much lower temporal resolution than in the case of syllables. On the other hand, although consonant parts contain a great deal of valuable information, with the exception of nasal sounds, most of them are characterized by considerably smaller volume levels than vowel sounds. Therefore, when using a new method of sound compression, careful consideration must be given to adjusting the level of every syllable. Specifically, in the case of vowel sounds, amplification must be controlled to prevent the loudness from going beyond a person's comfort zone, while, at the same time, the volume of consonants must be sufficiently amplified to enter that person's hearing range.

2.4.2   Enhancement Methods of Consonant Parts

The Japanese government launched a project to promote the development of digital hearing aids in 1995. Two specific goals of this project were to improve hearing aids by taking advantage of DSPs to find ways [4] to better compensate for the decrease in frequency resolution as well as the deterioration of temporal resolution that is characteristic of the sensory neural hearing impaired. Additional goals were to find better solutions for such problems as noise interference, narrow dynamic range and the recruitment phenomenon.

In this digital hearing model, the frequency axis of the lateral inhibitory function—the basis of signal processing—was represented by the horizontal axis in order to emphasize formants. In addition, decreasing the masking effect on the temporal axis was attempted by activating the lateral inhibitory operators on the temporal axis that corresponded to the temporal sensory characteristics, as will be discussed in Chap. 4. Therefore, the lateral inhibitory function was formed on the temporal axis and the code-reversed lateral inhibitory function (Fig. 2.10a) was convoluted into input signals in order to provide certain temporal differential effects. This method emphasizes excessive signals such as consonant sounds because the value increases before and after the level fluctuations of the signal envelop line. In order to suppress the peaks in vowel sounds, the input signal was multiplied by a function as shown in Fig. 2.10c. Also, as the input-output characteristics of Fig. 2.10d indicate, the maximum benefit was limited to 15 dB so that excessive amplification would not lead to errors in hearing. When the lateral inhibitory function code was reversed, two types were proposed: a symmetrical type shown in Fig. 2.10a, and an asymmetrical type in Fig. 2.10b. The relative effectiveness of the two types was then evaluated.

For this evaluation, an articulation test was conducted on 10 people with normal hearing by using vowel-consonant-vowel (VCV)voice sounds combined with white noise, where $S/N = -5$ dB. The articulation score results indicated a slight improvement, especially in the case of the asymmetrical lateral inhibitory function.

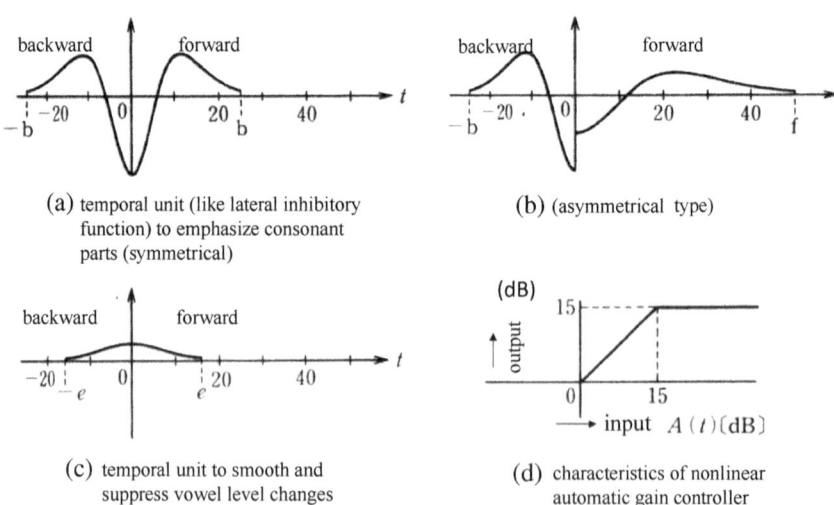

(a) temporal unit (like lateral inhibitory function) to emphasize consonant parts (symmetrical)

(b) (asymmetrical type)

(c) temporal unit to smooth and suppress vowel level changes

(d) characteristics of nonlinear automatic gain controller

Fig. 2.10 Time convolution methods to emphasize consonants and suppress vowel levels

However, the number of hearing errors for voiceless plosive sounds increased. Therefore, the average articulation score using all VCVs did not show a significant improvement.

As a case in point, one recurring problem is that consonant emphasis processing can also result in the unwanted emphasis of noise that resembles consonants. By the same token, the suppression of noise may also be accompanied by the undesirable effect of suppressing important elements of voice sounds. Nevertheless, the gain control functions including the emphasis of the weak consonant levels and the suppression of the loud vowel levels would be useful for some sensory neural hearing-impaired people.

2.4.3 Noise Reduction Methods

Many methods of noise reduction have been proposed and some of them may be useful for digital hearing aids. However, careful attention should be paid when adopting noise reduction methods for use in hearing aids. Specifically, there is always the pitfall that the signal itself might be misinterpreted as noise; thus, resulting in the loss of the original signal. This is because in many cases, for those on the receiving end of the sounds, it is difficult for them to know in advance if the input sounds are actually just noise or meaningful signals.

2.4.3.1 Reproduction of Voice Sounds Recorded on Old Wax Cylinders

As a case in point, the author and his colleagues were once asked to reproduce sounds recorded on 72 wax cylinders that had been found in Poland (Fig. 2.11) [5]. Specifically, the wax cylinders contained recorded songs and musical performances of Hokkaido's indigenous Ainu people. As the recordings were more than 100 years old, the surface of many of them was covered with mold, resulting in degraded quality. Thus, our primary objective was how to decrease the noise in order to make the actual voice sounds clearer. Specifically, we used several kinds of signal-processing algorithms in the hope that attaining a significant noise reduction might prove helpful in the development of digital hearing aids. A reproduction device that we developed is presented in Fig. 2.12 [6]. In the device, a laser beam [7] as well as a conventional needle sensor can detect sounds from the wax cylinders and then the noise components of the detected signals are decreased by following three signal processing methods.

In fact, the noise components of all produced sounds were decreased using GHA (Generalized Harmonic Analysis) method that Muraoka and his colleagues developed [8]. The author and his colleagues reproduced an Edison wax cylinder that contained a piece "Hungarian Dance No. 1" played and recorded by its composer "Johannes Brahms" himself in 1889. Muraoka and his colleagues

Fig. 2.11 One of wax cylinders (*right*) and its case (*left*) that polish anthropologist Piłsudski recorded more than 100 years ago

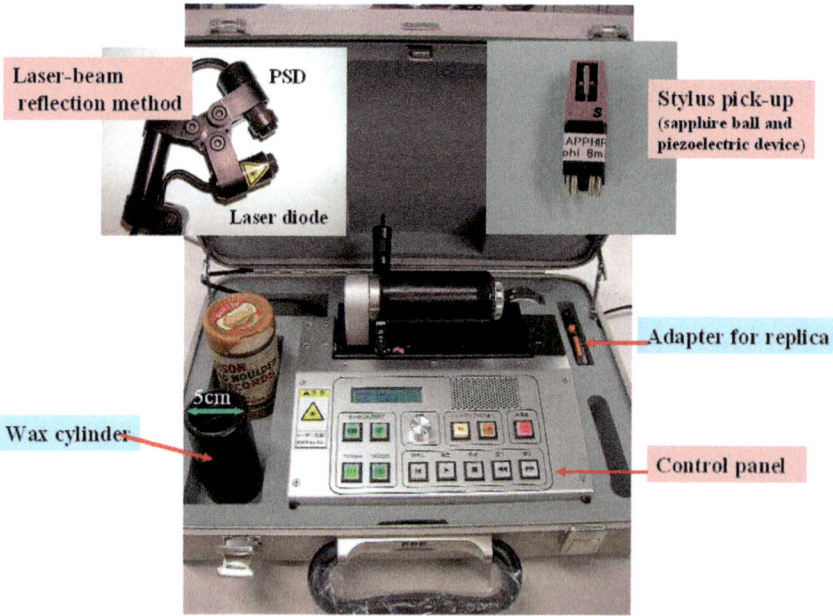

Fig. 2.12 Reproduction apparatus of the wax cylinders. Two methods were designed: laser beam reflection and pick-up made of Piezo-electric device

succeeded to decrease the noise components of the piece in 2013 [9]. Unfortunately, since the GHA method takes an enormous amount of time and it required skill, it is difficult to apply the GHA for the noise reduction method in digital hearing aids.

2.4.3.2   Spectral Subtraction Method

For the first attempt, the spectral subtraction method was used. With this method, if the quality of noise is known to some extent, the power spectrum of voice signals can be assumed by subtracting the short temporal power spectrum of noise from the short temporal power spectrum of the voice signals. Then, by squaring the root, the short temporal amplitude spectrum was calculated. Lastly, sounds that were close to the voice signals and within the voiceless section were used. This principle, as shown in Fig. 2.13, constitutes what is known as the spectral subtraction method.

As Fig. 2.13 indicates, when compared to the other methods, the noise reduction rate obtained from the spectral subtraction method was highest. However, that rate largely depends on the degree of subtraction. When too much was subtracted, for example, the sound quality decreased because the consonant sounds that had weak sound pressure were lacking. Conversely, when too little was subtracted, irritating noise elements remained. In short, only when approximate information regarding the particular noise is known, can this method be applied for use in hearing aids.

Fig. 2.13 Spectral subtraction method

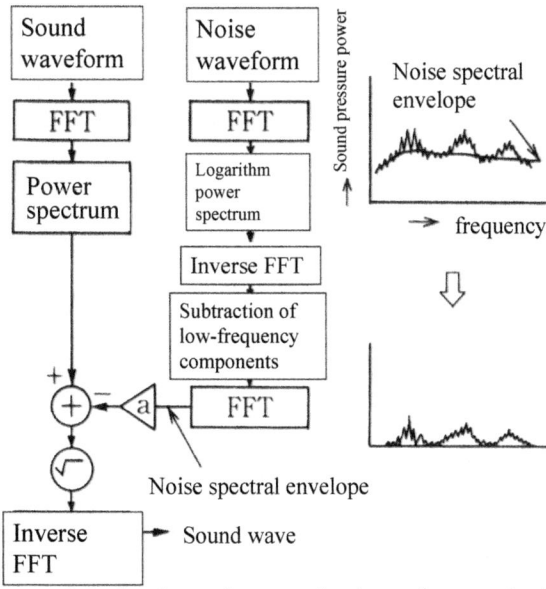

Mechanism of spectral subtraction method

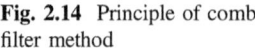

Fig. 2.14 Principle of comb filter method

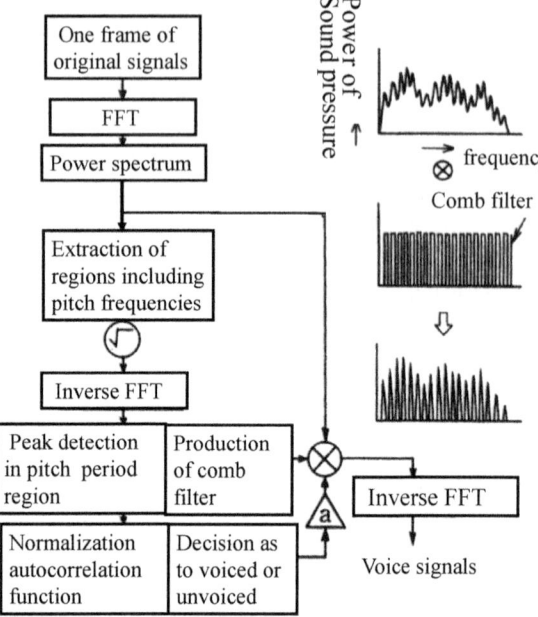

2.4.3.3   Comb Filter Method

For the second attempt at noise reduction, the comb filter method was chosen. As voiced sounds have periodicity, the spectrum possesses a line spectral structure consisting of a basic frequency and harmonics, as shown in Fig. 2.14. Therefore, noise can be decreased by leaving only this spectral part while suppressing the other parts. The accuracy of this method is determined by how accurately the pitches and distinguished voiced sounds can be detected from voiceless sounds. As the pitch detection method, the most orthodox autocorrelation function was used. As voiced sounds are quasi periodic waves, the basic frequency, namely "pitch frequency", is calculated from the autocorrelation function point that has the maximum level.

In distinguishing voiced from voiceless sounds, one hint is the correlation value. After determining this preprocess, the comb filter, the spectral figure of which is shown in the figure, was obtained from the calculated pitch period. The comb filter was used within the frame judged as voiced sounds, but not used within the frame judged as voiceless sounds.

In the case of the comb filter method, both the noise reduction rate and the sound quality improved. Moreover, in the case of the wax cylinders, this method was almost as effective as white noise. The spectra are cut off in certain places, with the remaining spectral portions being judged as voiced sounds. However, in the case of the wax cylinders, it so happened that the voiceless parts were either misjudged as voiced parts or pitch detection did not function properly. This occurred because noise elements themselves, which concentrate around 1.2 kHz, have certain cycles.

Therefore, although the noise decreased, it was accompanied by a reduction in sound quality. If this method is applied to hearing aids, it should be effective for those noises that do not contain any periodicity.

2.4.3.4   Autocorrelation (SPAC) Method

For the third and final attempt at using signal processors, speech processing by the autocorrelation method (SPAC) developed by Suzuki and his co-researchers [10], which uses autocorrelation functions to decrease noise, was chosen. In the case of white noise, the energy from the autocorrelation function concentrates around a zero time delay. Therefore, if the noise mixed in with voice sounds is close to white noise, by removing the waves around the zero time delay of the autocorrelation function, the subsequent decrease in noise power leaves sound elements that contain periodicity. In the case of voice sound waves, a short time autocorrelation function is used to calculate each frame, since signals change over time.

The actual process is indicated in Fig. 2.15. The shortcoming of this method is that there is no way to avoid the loss of non-periodic elements, such as frictions, since they resemble noise. Another problem is that the accuracy of this method is largely influenced by the accuracy of pitch detection.

Although noise reduction by the autocorrelation method is almost as effective as the spectral subtraction method in improving sound quality, it sometimes leads to a breakdown in the harmonics of voice sounds. It goes without saying that this results from a malfunction in pitch detection. In fact, this malfunction is distinctive when

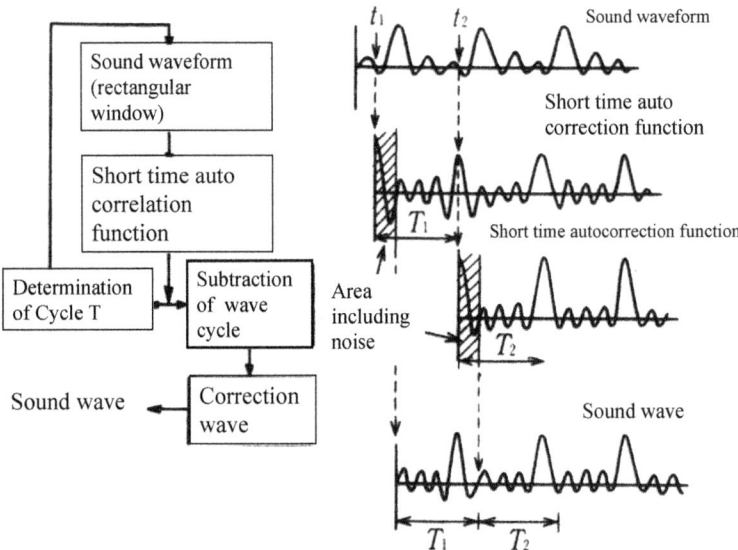

Fig. 2.15 Principle of speech processing by autocorrelation (SPAC)

the noise is periodic in nature, as in the case of the wax cylinders. Furthermore, since non-periodic elements such as frictions and plosives are also omitted during this process, it becomes even more difficult to hear consonant sounds.

2.4.3.5 Noise Cancellation Methods

For the sake of a high-speed processor, noise cancellation methods have been realized in various fields such as noise-cancellation headphones. Although this method was not used for the noise reduction of the wax cylinders, the author would like to add it as it is one of candidates for noise reduction in digital hearing aids. As shown in Fig. 2.16, even when noise signals (external sound) are added to the audio signal that listeners really want to hear, the audio signal may be clearly heard by releasing other sound pressure waves with the same amplitude but with an opposite or inverted phase (antiphase). This effect is also called "phase cancellation." In short, the effectiveness of the processing method depends on various factors; including pitch detection, accuracy in judging voiced from voiceless sounds, the level of polarization and periodicity of noise elements, and on how close the eliminated noise elements are to the noise elements that have been added.

   Although there were various kinds of noise reduction methods for hearing aids, only a few of them clearly surpassed the efficiency of conventional sound pressure amplification-type analogue hearing aids. This is because there are only a few known methods that can effectively examine the information-processing function of the auditory sense in order to device ways to improve its function without causing

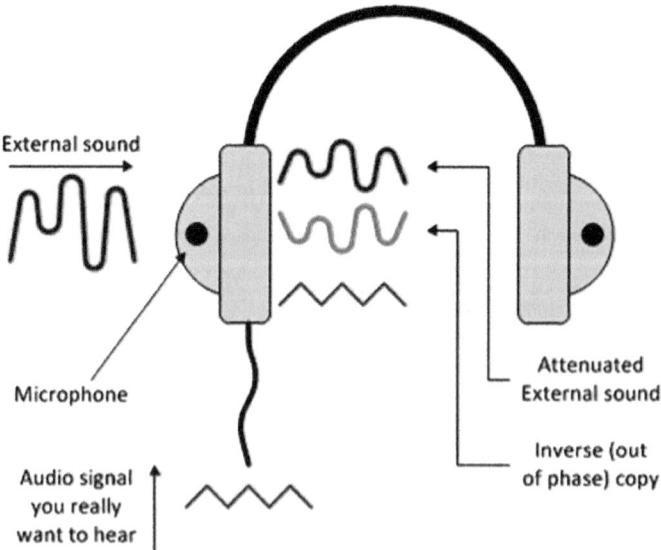

Fig. 2.16 Mechanism of noise-cancellation headphones

any damage. For digital hearing aids to truly serve as an effective hearing method, it is most essential to thoroughly examine how information processing in people with normal hearing differs from that of the hearing impaired. Present-day research on digital hearing aids has tried to adequately address this problem.

2.4.4 Speech Speed and Intonation Conversion Methods

When one talks to the elderly or the hearing impaired, one has an unconscious tendency to speak more *slowly* and *clearly*. Moreover, such people actually wish to be spoken to in such a manner. Generally, in the case of the elderly hearing impaired, the hearing defect is thought to result from hearing loss in the auditory peripheral nervous system combined with a deterioration in both short-term memory as well as in the overall ability to process language.

2.4.4.1 Slow Down Speech Speed Method

In regard to this matter, there has been an attempt to assist people who experience hearing loss by slowing down the speed of speech by means of signal processing. Such a hearing device, equipped with a converter to *slow down* speech, was designed by Nejime and his colleagues in Hitachi, LTD. Their method was realized as a pocket-sized device from Hitachi to improve its practical use in 1994 [11], as shown in Fig. 2.17a.

The method is called a "speech rate converter (SRC) hearing device," and it was significantly different from usual hearing aids. Both devices employed a method of extracting certain waves from the repeating waves of primarily vowel sounds, after which those sounds were repeated and lengthened. Their method originally consisted of a hearing device that enabled a user to detect voice sounds obtained from one source—similar to the voice sounds heard from a radio news broadcast—while continuously slowing down the speed of the sounds. The principle of speech speed conversion as well as the effects of this method have been shown by describing a pocket-sized device, which may serve as a representative example that corresponds most clearly to the image people have of hearing aids. For the frequency-processing characteristics, the software module shown in the figure divides voice sounds into four bands, so as to amplify the amplitudes and to compress the dynamic range of each band.

When using this method, the intervals at which time the power of each sound frame exceeds the threshold are fixed as points where the sounds duration is extended, as shown in Fig. 2.17b. Moreover, this sound wave extension process is only used for the sound frame data contained in those specific intervals. Based on conventional subjective evaluations, when the soundless period of a voice is too long, it sounds unnatural to the listener, whereas when soundless periods exceeding one second are compressed into a duration of one second, it seems that the

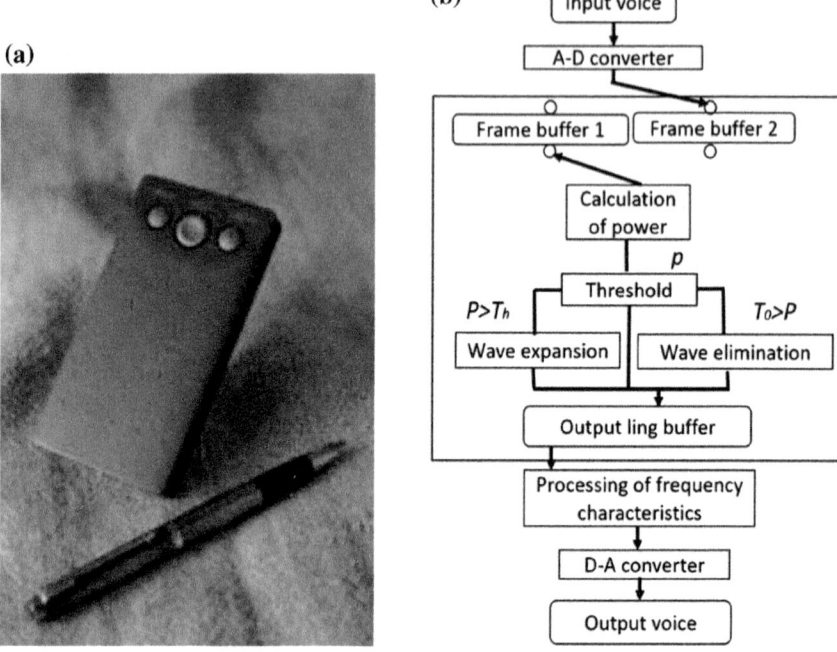

Fig. 2.17 **a** Appearance of portable digital hearing aid with speech rate conversion function, **b** decision algorithm of durations for expansion and elimination

naturalness of the sound is maintained. Therefore, this method consists of detecting soundless periods based on their threshold level in order to eliminate those whose duration exceeds one second so that they will not be sent to the output buffer.

For the frames that are stretched by this threshold level process, the wave is processed for each pitch in the temporal zone and the temporal elongation of sounds is conducted without changing the pitch. A well-known method of extending the duration of sound waves for each pitch is time domain harmonic scaling (TDHS), a process originally developed for compressing information. Some methods have been proposed to improve TDHS. Without going into detail, one of these proposed improvements (Fig. 2.18) involves the principle of wave stretching. When using this method, three stretch rates (e) are obtained: $e = 1.50$ ($2 \rightarrow 3$ pitches), $e = 1.33$ ($3 \rightarrow 4$ pitches), and $e = 1.25$ ($4 \rightarrow 5$ pitches).

After evaluating the method by the author and Nejime, they are convinced that this method is an effective aid to enable the sensory neural hearing impaired to hear sentences. Figure 2.19 shows the relationship between the rate of hearing errors (vertical axis) and the sound elongation rate (the horizontal axis) for 10 elderly hearing-impaired people who heard meaningless sentences consisting of four words. It is clear that the rate of hearing errors decreased for all subjects when they were able to hear the sounds more *slowly*.

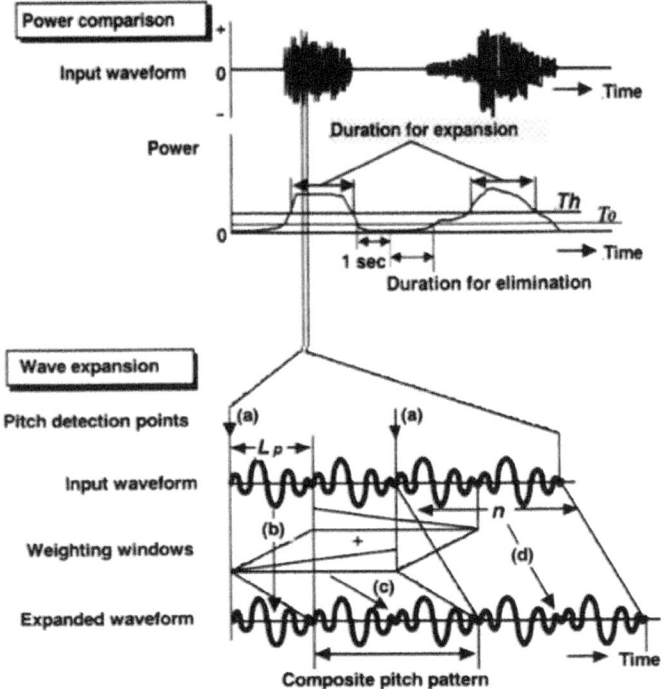

Fig. 2.18 *Upper* Decision of duration for expansion; *lower* wave expansion method

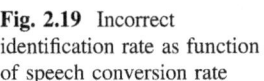

Fig. 2.19 Incorrect identification rate as function of speech conversion rate

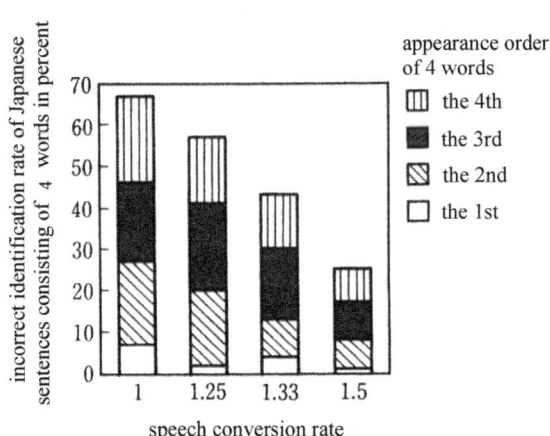

Figure 2.20 shows the relationship between temporal resolution (horizontal axis) based on the perception of two successive sounds and the improvement in error rate (vertical axis). It shows that the poorer the subject's temporal resolution, the greater their rate of improvement.

Fig. 2.20 Improvement of
identification rate as function
of auditory time resolution

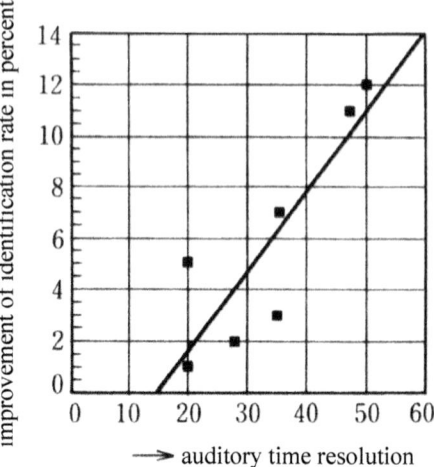

As temporal resolution of the auditory sense generally declines with age, it is assumed that hearing aids that enable elderly hearing-impaired people to hear at a slower rate would provide valuable assistance. Although our evaluation was conducted on the hearing impaired, hearing-impaired people who also have a serious brain defect, such as Wernicke's aphasia, run a greater risk of developing a decline in temporal resolution. In view of this situation, it is necessary to examine whether the speech speed conversion method could also be effective for them.

However, this SRC (speech rate conversion) method would not be practical for hearing aids unless it were fully automatic and in real time. Imai who belongs to NHK (Japanese Broadcast Co.LTD) and his colleagues proposed "adaptive SRC" that retains or prolongs only speech parts that are likely to have a significant effect on listening comprehension [12]. When saying something important, people raise the pitch or power of their voice especially in Japanese. On the other hand, if both the power and the pitch of a voice are relatively low in a particular section of a continuous utterance, it may be removed or shortened because it is expected to be difficult to hear in original speeches. From these viewpoints, they developed the algorithm used in their adaptive SRC method. The algorithm observes time fluctuations of the power and F0 (fundamental frequency) of the speech signal and deletes sections in which they fall below a certain threshold so that the duration of the processed sentence can be the same as the original sentence. Figure 2.21 schematically shows the design concept of the adaptive SRC.

Although the detail of the effectiveness of the adaptive SRC is omitted, they have proved that it is indeed useful for the elderly to listen a radio broadcasting. Furthermore, the adaptive SRC method can be applied by changing the thresholds regarding the speech power and F0 for "ultrafast listening" that is equivalent to skimming. Therefore, this algorithm will help not only elderly people to enjoy radio broadcasting but also visually impaired people because almost all of their information is obtained from speech as mentioned in Chap. 7.

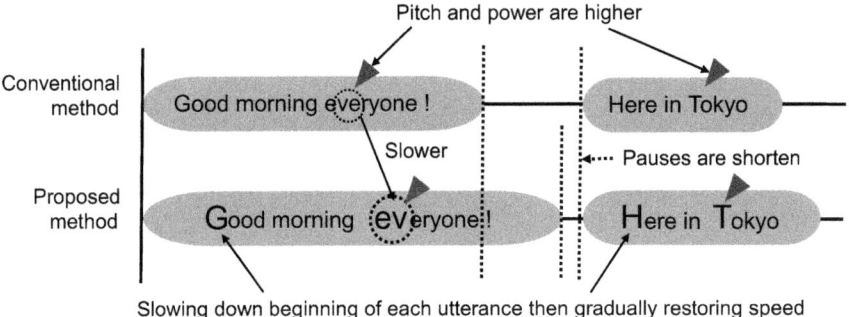

Fig. 2.21 Design concept of adaptive SRC technology

2.4.4.2   Pitch Change Enhancement Method

On the other hand, when clearly spoken sounds are analyzed, it is commonly observed that intonation is emphasized. Recently, research has been undertaken that focuses on sound recognition and the extraction of emotional sounds based on phonetic data such as intonation accent and rhythm. It has also been reported that a fairly high recognition rate can be obtained through the limited use of lip reading combined with the phonetic method. It is therefore easy to imagine that phonetic information plays an important role in the perception of voice sounds.

In the Japanese language, there are numerous words whose meanings vary when the same basic pronunciation is accompanied with a different intonation. One typical example, as shown in the *left* side of Fig. 2.22, shows what occurs when the intonation pattern of the word "*ame*"—meaning either "rain" or "candy"—is reversed.

The author and his colleagues preliminary investigated the role of intonation on the discrimination of the two words [13]. The *right* side of Fig. 2.22 indicates the

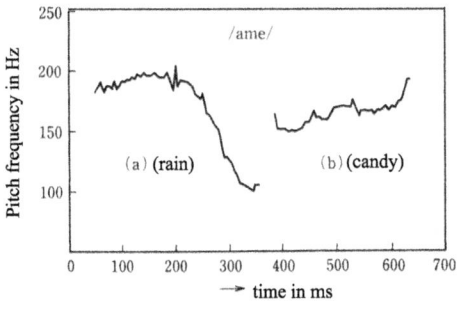
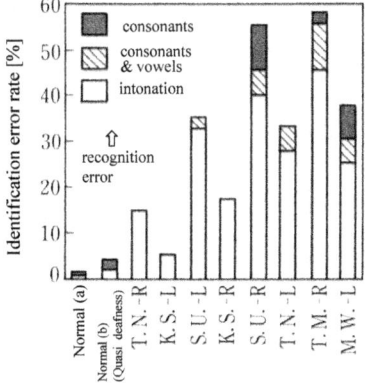

Fig. 2.22 *Left* Pitch frequency patterns for /ame/ with decreasing pitch frequency (rain in English) and /ame/ with increasing pitch (candy in English), *Right* rate of hearing errors between *ame* (rain or candy) and *kame* (turtle or pot), whose pronunciations are the same but whose intonations vary

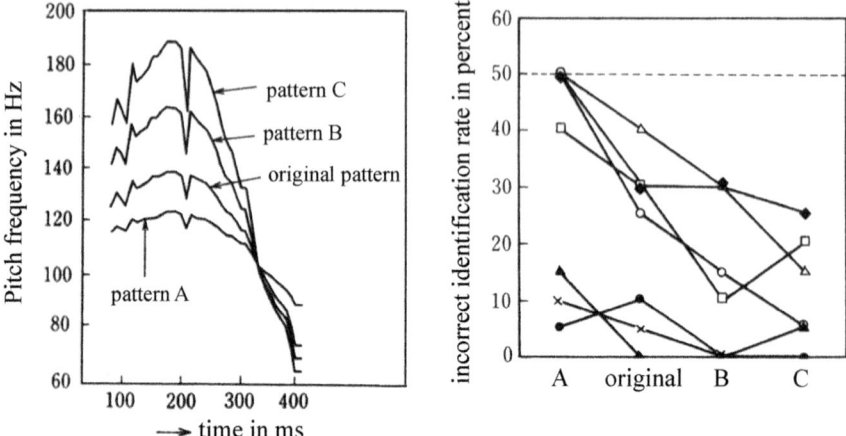

Fig. 2.23 *Left* Pitch change enhancement for /ame/, *Right* incorrect identification rate of pitch enhanced voices by sensory neural hearing-impaired people

rate of hearing errors that occurred when eight people with moderate sensory neural hearing-impaired and two adults with normal hearing attempted to distinguish between *ame* (rain or candy) and *kame* (turtle or pot) whose pronunciations are the same but vary in intonation. Where, *kame* with decreasing pitch frequency and *kame* with increasing pitch frequency mean "turtle" and "pot" in English, respectively. All of those subjects used standard intonation that was free of any dialect. The rate of hearing errors of the subjects with normal hearing was less than 5% as shown in two bar graphs, (a) and (b), on the left side of the figure. On the other hand, the rate of hearing errors was around 33% (max: 58%, min: 6%) in the case of the subjects with hearing-impaired. Especially, it was clear that they are hard to distinguish a clear variance in intonation (white portion of the bars).

   Therefore, as the right figure of Fig. 2.23 indicates, we calculated the rate of hearing errors resulting from the utterance of words that required an intonation emphasis as shown in the left figure of Fig. 2.23. It was found that even when the subjects heard those words at random, the rate of hearing errors decreased as a result of their intonations.

   From the results of this examination, very little correlation was found among hearing level, hearing type, and the intonation distinction ability of the sensory neural hearing impaired. A deterioration in the ability to distinguish intonation cannot be explained merely by a deterioration in peripheral auditory ability, which can be seen in an audiogram. This innovative method of enabling people to hear more *slowly* and *clearly* offers yet another potential for the improvement of hearing aids.

   Lu and the author found that the enhancement of tone contrast in Mandarin may result in improving the speech reception of hearing-impaired listeners. In summary, the present results suggest that some perceptual advantages might be gained by enhancing tone contrast for hearing-impaired Chinese. It is known that the pitch

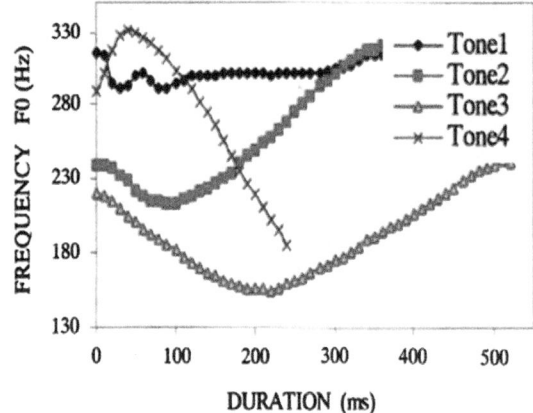

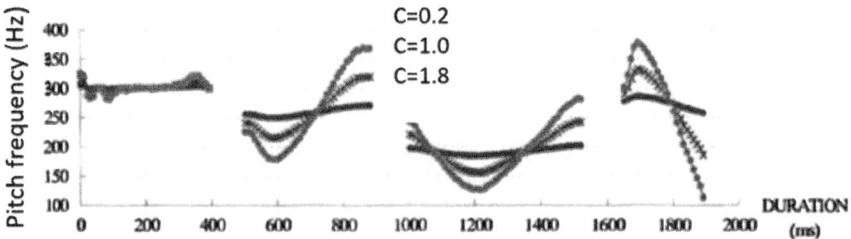

Fig. 2.24 Pitch variation in Mandarin Chinese /ma/ (*upper*) and Contrast enhancement (*lower*)

variation in Chinese is more striking than in other languages. Mandarin has four lexical tones that include a high-level tone (tone1), a mid-rising tone (tone2), a low-falling-rising tone (tone3), and a high-falling tone (tone4).

These four tones are distinguished by relative pitch changes such as the position of the turning point, as well as the rise or descent of each syllable. In a Mandarin word, the difference in the tones creates different meanings even if the phonemes have the same composition. An example of the four tones for one speaker is shown in the top figure of Fig. 2.24. Based on the characteristics of Mandarin itself, as well as the hearing properties of hearing-impaired individuals, some form of signal processing seems warranted for Chinese who have difficulty in extracting F0 cues from natural speech.

Lu's study concentrated on a tone-modification processing method to improve the hearing of hearing-impaired people [14]. The tone contrast—defined as the difference in frequency between peaks and valleys of a tone curve—was exaggerated to produce a tone modification in which a greater-than-normal proportion of the tone contour would be more easily distinguished. The modified fundamental frequency of each analysis frame in the new tone contour was computed using the following relation:

$$FO_{new}(t) = C \times \left[FO_{orig}(t) - FO_{mean} \right] + FO_{mean} \qquad (2.1)$$

where FO_{new} is the modified pitch frequency in an analysis frame, FO_{mean} is the average fundamental frequency level of overall speech signals, and C is a contrast modification factor (hereafter referred to as "contrast weight").

When the contrast weight C was 1.0, the target tone contour was the same as the original tone. Contrast weights greater than 1.0 produce contrast enhancement, whereas contrast weights less than 1.0 produce contrast reduction. In the extreme case of reduction (C = 0.0), all features of the original tone contour would be lost in the target tone contour and would be flat at the average pitch. In order to insure the quality of speech sounds, the range of modified pitch was restricted between 70 and 400 Hz. Contrast weights of 0.0, 0.4, 0.6, 1.0, 1.5, 1.8, 2.0 were used for the modified stimuli in the experiment. Weights of 0.0, 0.4, 0.6 were used as the contrast-reduced stimuli, while weights of 1.5, 1.8, 2.0 were used as the contrast-enhanced stimuli. As an example, the modified tone2 is illustrated in the bottom figures of Fig. 2.24.

Figure 2.25 shows the mean percentage of correct identification for each level of contrast weight, with the percentage of correct identification increasing from reduced to normal and from normal to enhance contrast. From this figure, it can be seen that tone-enhanced speech results in a moderate increase in the percentage of correct word identification compared with unmodified speech. In contrast, tone-reduced speech generally reduced the percentage of correct word identification. The observed contrast effects were also apparent in speech intelligibility of listeners with moderate (average hearing level is around 40 dB) and severe (average hearing level is more than 60 dB) sensory neural hearing loss.

By using a speech-modification algorithm that manipulates the speech pitch, it is hypothesized that the processed speech is effective in enhancing the language function of the CNS in hearing-impaired subjects.

Fig. 2.25 Percent of correct tone identification for each subject group in each contrast condition

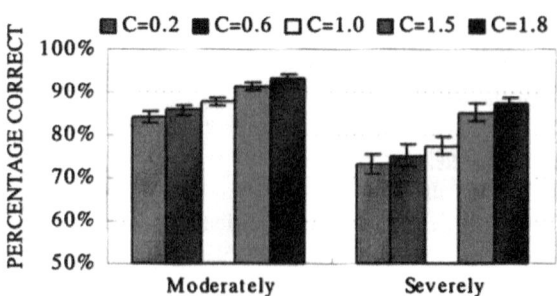

2.5 Artificial Middle Ears

Some common problems of conventional hearing aids are that users might feel aggravated or uncomfortable while wearing them. Moreover, if the device looks unusual, they may even be afraid of attracting undue attention. Another problem is that when they are unable to wear such devices, during such activities as bathing, sleeping or doing strenuous exercise, they cannot hear important warning sounds. In addition, there remains one technical problem to be solved, namely that increasing the volume of a hearing aid may cause sound distortions or howling.

Artificial middle ears were developed to solve the above-mentioned problems. However, such devices can only be used if the hearing defects are limited to the sound-conducting system and the users have no serious defects involving the auditory nerves in an area deeper than the inner ear.

Artificial middle ears utilize a method of converting sound signals into mechanical vibrations and conveying those vibrations directly to the auditory ossicles or inner ears. Artificial middle ears can be categorized as a kind of artificial organ since they involve the implantation of a transducer, an amplifier, and a power source, etc., into a human body. In general, there are basically two kinds of implants: total implants, which, as the name implies, involve implanting the entire device into a body; and partial implants, where only part of the device is implanted.

The totally implanted type can be used while bathing or swimming and does not require maintenance. However, as the life of the power source is limited, surgery is necessary each time the batteries need changing. One type of partial implant consists of implanting a charged battery into the ear and electromagnetically stimulating the device through a coil, as shown in Fig. 2.26a [15]. In this case, although periodic recharging is still necessary, it is possible to implant a smaller battery than that used in a total implant. However, as in the case of the total implant method, this approach also includes implanting the microphone and the amplifier.

For another kind of partial implant, only a Giant Magnetostrictive Material (GMM) that can produce a strong vibration is located inside the body where it conveys the signals (Fig. 2.26b) [16]. Consequently, the microphone and amplifier are worn outside of the body. Unlike the total implant, it is possible to adjust the level of amplification or frequency characteristics outside of the body. However, as in the case of conventional hearing aids, this device may prove somewhat troublesome or annoying to the user from the point of view of appearance.

When using such implants, however, the biggest obstacle to overcome is their housing rather than the development of electronic technology. That is to say, careful consideration must be made in choosing an exterior material for the device that has no effect on living tissue. Specifically, the primary concern is that the exterior material must both prevent body fluids from entering the device and ensure that there is no leakage from the device. As vibrators lie at the heart of a properly functioning artificial middle ear or implanted bone conduction hearing aids, one must be careful to ensure that the vibrators and the bone are securely joined and that they will function correctly over a long period of time. For this reason,

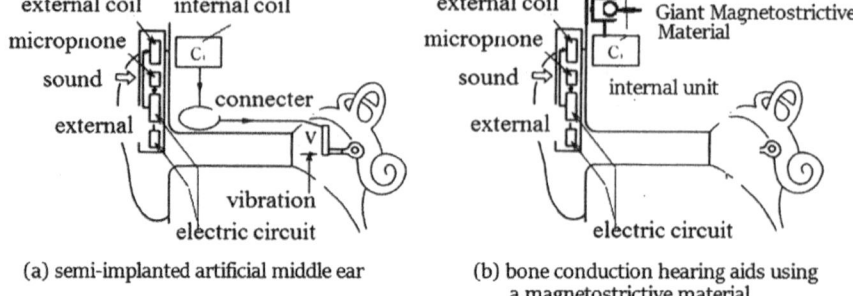

(a) semi-implanted artificial middle ear   (b) bone conduction hearing aids using
                                                      a magnetostrictive material

Fig. 2.26 **a** Schematic representation of semi-implanted artificial middle ear using a piezoelectric vibrator, **b** bone conduction hearing aids using giant magnetostrictive material

dental-purpose adhesive—a highly suitable bonding agent for living tissue—is used to join the vibrators to the stapes, while the supporting part is fixed to the temporal bone with screws.

Until 1993, approximately 60 subjects received benefit from artificial middle ears shown in figure (a), resulting in their approval by the Japanese government as a highly advanced form of medical technology. However, recently, the implanted bone conduction hearing aids shown in figure (b) has been replaced instead of the artificial middle ears. As mentioned in Chap. 3, these implant methods and housing technologies of the artificial middle eras are useful to design cochlear implants.

2.6 Design of Earthquake Early Warning Chime Based on Auditory Characteristics [17]

The Great East Japan earthquake occurred and an earthquake early warning chime, which the author composed in 2007, repeatedly rang all over Japan. The earthquake that struck on March 11th, 2011, caused an unprecedented disaster. It created approximately 20,000,000 victims—including missing people—and destroyed more than 270,000 homes. The number of aftershocks was extraordinary; more than 500 were recorded in the three months after the main strike due to the massive earthquake. Topic is shifted slightly but the author would like to lightly touch upon the chimes sound, since he designed the sound in consideration of the hearing characteristics so that it may be heard even in the hard of hearing.

There two kinds of tremors; P waves, small pitch (preliminary tremors) and S waves, main tremors (principal shock) as shown in Fig. 2.27. These two occur at the same time, however P waves have a faster propagation speed. Therefore, the further from the focus, the bigger the time lapse between P and S waves. Everyone should have experienced that the bigger tremors come after the first shock. The earthquake early warning is sent to the public as soon as possible when the system

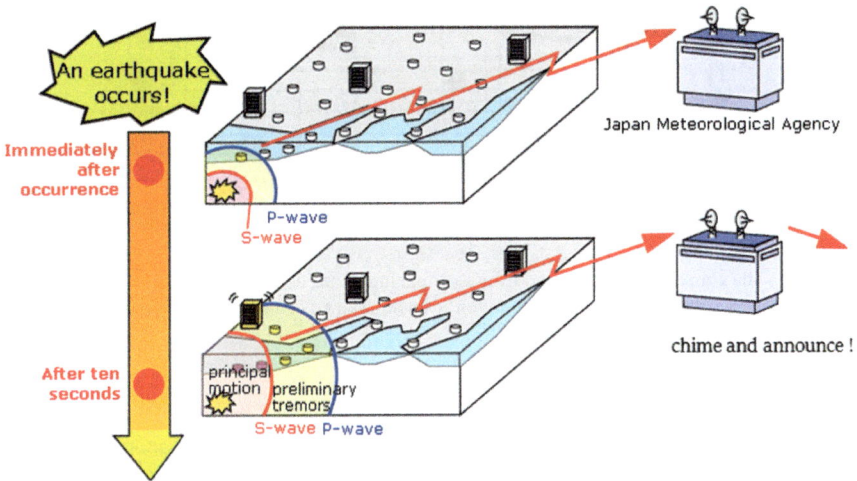

Fig. 2.27 Earthquake early warning system [from Home page of The Japan Meteorological Agency (JMA)]

senses the P waves and the prospective earthquake intensity is above a low 5, before the S waves arrives.

The author was asked to work on the chime sound by NHK (Japan's national public broadcasting organization) since he specializes in assistive technology which is related to hearing and visual impairment. He proposed the ideas for the earthquake early warning, there are five important conditions that have to be satisfied. (1) The sound has to attract attention. (2) The sound has to encourage people to act. (3) The sound has to be different to any kind of chimes sound or alarms that already exist. (4) The sound has to be a decent tone; not too uncomfortable or comfortable and not too optimistic or gloomy. (5) The sound has to be heard by as many people who have hearing problems as possible.

After examining a number of sound and music, the author selected the opening chord of "Vivace," the third movement of "Sinfonia Tapkaara" created by the composer Akira Ifukube. The composer is the author's uncle and is well known that he composed the theme music of "Godzilla". The reason why the author focused on the music, not buzzer or alarm sounds was that he wanted the sound to have some kind of message. Actually, the first harmonic chord has "tension note" that gives tense feelings. Furthermore, the author arranged the chord into five types of the chime all of which are rearranged into "arpeggio" that is a kind of a frequency modulated tone (FM tone).

Then, he conducted an experiment with 19 participants including 10 people who had hearing impairments. A total of 19 participants; 6 children (3 of which had hearing difficulties), and 13 adults (5 of which had hearing difficulties) in this

Fig. 2.28 The chimes sound arranged from the opening chord of "*Vivace*," the third movement of "Sinfonia Tapkaara" composed by the author's uncle Akira Ifukube who is well known that he composed the theme music of "Godzilla"

experiment. The experiment was performed under typical noises in our daily lives such as multi-conversational noise (conversations between many people), musical noise, traffic noise, and the noise of amusement arcades. Participants were asked to listen to 5 chimes sounds in pairs, and answer the questions such as "Which chimes sound expresses emergency more?" After the verification experiment, the chimes sound was finally selected as shown in Fig. 2.28. This earthquake early warning system has been in place since August 1st, 2006.

Now, Japanese people are familiar with the alarm warning: "This is an earthquake early warning. Please prepare for powerful tremors." followed by two sets of chimes. Every time we heard this warning, we crawled under the tables and were aware of what was overhead and of the fire situation. This alarm sounded repeatedly on TVs and radios each time a large aftershock was detected, and these chimes was clearly imprinted into our brains.

2.7 The Future of Hearing Aids

Apart from conductive deafness, in many cases the cause of a decrease in hearing ability is not only due to malfunctions involving either the nerves leading to the inner ears or the inner ears themselves, but also to a deterioration in the function of the brain's language centers. More specifically, some defects in the brain's language centers can lead to aphasia, which in turn, results in hearing loss. Unfortunately, the hearing aids presently available are ineffective in cases involving aphasia. An important point worth mentioning, however, is the belief among researchers that it is easier for aphasics to hear when they are spoken to slowly. In future, the whole concept of hearing aids will have a much wider application, including the important goal among researchers of considering hearing aids as actually comprising a means of maintaining the proper functioning of the various language centers of the brain.

References

1. T. Sone, Y. Suzuki, K. Ozawa, F. Aasano, Information of loudness in aural communication. Interdisc. Inform. Sci. **1**(1), 51–66 (1994)
2. T. Imamura, T. Ifukube, J. Matsushima, H. Syoji, Comparison of combination tone perception between normal hearing listeners and the hearing impaired by using masking technique. Audiol. Jpn. **38**(1), 66–76 (1995)
3. F. Asano, Y. Suzuki, T. Sone, S. Kakehata, M. Satake, K. Ohyama, T. Kobayashi, T. Takasaka, A digital hearing aid that compensates loudness for sensorineural impaired listeners. Acoust. Speech, Signal Process. ICASSP-91, 3625–3628 (1991)
4. H. Levitt, Digital hearing aids: a tutorial review. J. Rehabil. Res. Dev. **24**(4), 7–20 (1987)
5. T. Ifukube, T. Kawashima, T. Asakura, New methods of sound reproduction from old wax cylinders. J. Acoust. Soc. Am. **85**(4), 1759–1766 (1989)
6. T. Ifukube, Y. Shimizu, A portable record player for wax cylinders using both laser-beam reflection and stylus methods. in, Proceedings of the AES 121st Convention, San Francisco, CA, USA, The Journal of the Audio Engineering Society 339 (2006)
7. T. Iwai, T. Kawashima, T. Ifukube, T. Asakura, Sound reproduction of old wax cylinders by using laser reflection method. Appl. Opt. **25**(5), 597–604 (1986)
8. Muraoka T, Miura T, Ochiai D, Ifukube T (2007) Theory of short-time generalized harmonic analysis (SGHA) and its fundamental characteristics. in, Audio Engineering Society, Convention Paper at the 123rd convention, New York
9. T. Muraoka (2015) Tribute to Mr. N. Nozawa The Eminent Authority of 78 rpm Shellac Records. Level No.: KK-Ushi, CD No.: KDHKO-50
10. J. Suzuki, Speech processing by splicing of autocorrelation function. in, Acoustics, Speech, and Signal Processing, IEEE International Conference on ICASSP'76, vol. 1 (1976) pp. 713–716
11. Y. Nejime, T. Aritsuka, T. Imamura, T. Ifukube, J. Matsushima, A portable digital speech—rate converter and its evaluation by hearing—impaired listeners: 1994 International Conference on Spoken Language Procession (ICSLP94), 32, 29 (1994), pp. 2055–2058
12. A. Imai, N. Tazawa, T. Takagi, T. Tanaka, T. Ifukube, A new touchscreen application to retrieve speech information efficiently. IEEE Transac. Consum. Electron. **59**(1) (2013)
13. T. Ifukube, T. Hokimoto, J. Matsushima, Y. Nejime, A digital hearing aid having a function of intonation emphasis. in, Proceedings on IEEE 17th Annual Conference, Engineering in Medicine and Biology Society, vol. 2 (1995), 1605–1606
14. J. Lu, N. Uemi, G. Li, T. Ifukube, Tone enhancement in Mandarin speech for listeners with hearing impairment. IEICE Trans. Inf. Syst. **E84-D**(5), 651–661 (2001)
15. J. Suzuki, K. Kodera, N. Yanagihara, Middle ear implant for humans. J. Acta Oto-Laryngol. **99**, 3–4 (1985)
16. Y. Sakai, S. Karino, K. Kaga, Bone-conducted auditory brainstem-evoked responses and skull vibratory velocity measurement in rats at frequencies of 0.5–30 kHz with a new giant magnetostrictive bone conduction transducer. Acta Otolaryngol. **126**(9), 926–933 (2006)
17. http://www.et97.com/view/1659272.htm (2016.12.10)

Chapter 3
Functional Electrical Stimulation to Auditory Nerves

Abstract In this chapter, first, the author discusses functional electrical stimulation (FES) of the auditory nerves using "cochlear implants", especially the history, principle, effects, and design concept of cochlear implants. Next, the author shows the recent progress in cochlear implants as well as auditory brainstem implants and secondary effects such as tinnitus suppression achieved by electrical stimulation. Finally, the author introduces "artificial retina" that has been served by the success of cochlear implants.

3.1 Functional Electrical Stimulation

Many of the sensory neural hearing impaired have merely lost the function of the hair cells in their inner ears, while their auditory nerves leading to the CNS remain intact. For this kind of hearing impairment, cochlear implants are an effective form of assistive technology. Specifically, such cochlear implants—made possible through a process known as functional electrical stimulation (FES)—convey sound information to the CNS by way of electrical stimulation to the surviving auditory nerves.

In general, the aim of FES is to design an effective interface between electronic equipment and a human being. One example of this method is to utilize electronic equipment to stimulate the surviving nerves and muscles connected to the limbs of people whose motor nerves have been severed, thereby resulting in a loss of mobility to the limbs. In this case, the users must naturally be able to control electronic apparatus simply by imposing their will.

Using a human hand as an illustrative example (see Fig. 3.1), electronic equipment first receives a signal that the hand has touched something; next, the signal provides electrical stimulation to the nerves that control the hand's movement. As a result, the hand may, for example, lift a glass up to a person's mouth, thereby allowing the person to drink. Although FES is mainly applied to those who have lost the movement of their arms or legs, it is particularly effective for patients whose nerves connected to the spine C5/C6 have been severed, thus impairing the

© Springer International Publishing AG 2017   77
T. Ifukube, *Sound-Based Assistive Technology*,
DOI 10.1007/978-3-319-47997-2_3

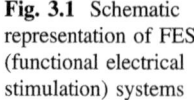

Fig. 3.1 Schematic representation of FES (functional electrical stimulation) systems

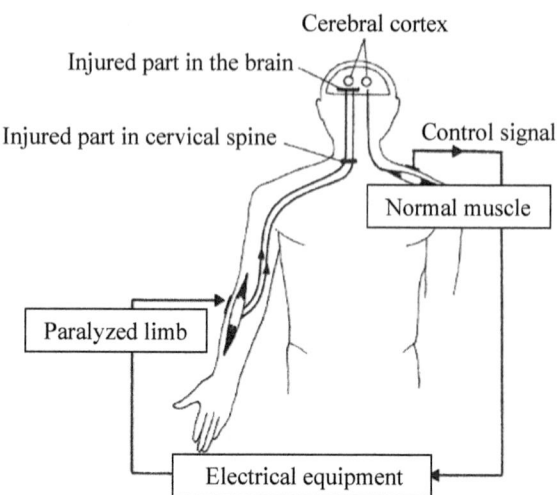

nervous system. FES research originated from the concept that by applying electrical stimulation to nerves specific to a desired movement, that impulse would be relayed to the peripheral nerves, thus resulting in movement of the limb.

Cochlear implants can also be categorized as a form of FES. Specifically, by providing electrical stimulation to surviving auditory nerves, the implants compensate for the impaired auditory sense in order to convey speech information to the CNS. They can also be categorized as artificial organs because part of the device is actually implanted into a person's body. However, the specific goal of cochlear implants differs somewhat from other forms of FES, the purpose of which is to enable a person to regain some movement or motor function. Specifically, FES to rebuild paralyzed limbs stimulates the efferent nerves, whereas cochlear implants provide information to the afferent nerves that lead to the CNS. Moreover, as cochlear implants are mainly used as an aid to communication between people, they offer real potential for the enhancement of social and intellectual life.

Speech information is conveyed to the brain's language center, where language concepts are formed. Accordingly, when the auditory function is damaged, the function of the brain's language cortex also changes. In particular, the effectiveness of cochlear implants differs greatly depending on whether the impairment occurred before or after a person's sensory and language concepts were formed. This is the most complicated and difficult problem to overcome when using cochlear implants during rehabilitation. In short, the development of cochlear implants is not simply a matter of imitating the function of the auditory organs by using a signal processor to convey information to the auditory sense.

However, even if there are some limitations to the application of sound-discrimination mechanisms or highly developed sound recognition functions that have been already formed in the auditory center, it is still more advantageous to make use of them than to compensate by attempting to develop these functions from the beginning by training the other senses. Moreover, as the brain possesses an excellent plasticity function, there is always the possibility of forming new concepts through learning, even if the information received is not always reliable. At this point, it is worth considering what has been described thus far, in order to further improve on the development of cochlear implants and their application to rehabilitation methods.

The idea of providing functional electrical stimulation to the auditory sense has existed for some time. It is said that the first person to notice that electrical stimulation applied to the auditory sense resulted in a sensation was Volta (Alessandro Volta, 1745–1827)—the same researcher who invented batteries. Specifically, he made the claim that he could hear sounds when holding a battery to his ear. In 1957, Djourno reported that although the hearing impaired were able to recognize the pitch of a voice when the auditory sense inside the inner ear was electrically stimulated, they were incapable of recognizing speech [1]. On the other hand, in 1964, Simmons (Francis Blair Simmons, 1930–1998) conducted an experiment that involved inserting six electrodes in the eighth cranial nerve that is connected to the inner ear. Through electrical stimulation, he found that the nerves stimulated by the electrodes combined with the pulse frequency provided by those nerves determined the pitch height of the perceived sounds [2].

As previously mentioned, however, unlike functional electrical stimulation applied to the limbs, the same approach to the auditory sense involves greater complexities. Although electrical stimulation of the peripheral nerve transmits an impulse to the auditory area, that area consists of many parts that are as yet to be fully understood. This fact alone sheds some light on the difficult task of how to best deal with speech signals. Nevertheless, research in the early 1980s into the most effective method of providing electrical stimulation to the nerves gradually helped clarify some of the problems involved, thereby yielding progress sufficient enough to enable the hearing impaired to finally use cochlear implants in a practical way.

3.2 Firing Process of Nerves and Electrical Stimulation Methods

3.2.1 Mechanism of Neural Firing

Here we will briefly describe the mechanism whereby neural cells are fired through electrical stimulation. Despite there being various theories regarding the process of how cells were initially formed, it is known with certainty that cells were born in seawater. Although seawater contains various ions—including sodium (Na^+), potassium (K^+) and chlorine (Cl^-), Na^+ is particularly abundant. Specifically, the relative proportion of Na^+, K^+ and Cl^- in seawater is 142, 5 and 103, respectively.

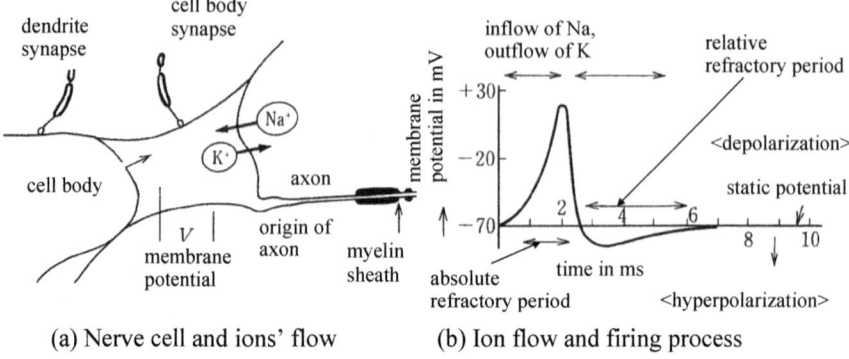

(a) Nerve cell and ions' flow   (b) Ion flow and firing process

Fig. 3.2 Structure of nerve cell and impulse production process

However, in the case of cells, the composition is quite the reverse, with the proportion being 10 (Na^+), 140 (K^+), and 7 (Cl^-).

As the component ratio inside cells is the opposite to what is found outside of cells, one would assume that Na^+ and Cl^- would naturally flow into cells whereas K^+ would flow out of the cells, thus resulting in cellular breakdown. However, the cell membrane prevents this from occurring. As there is a much greater abundance of Na^+ in seawater than that found inside cells, there is a natural tendency for Na^+ to attempt to enter the cells, as shown in Fig. 3.2a. However, under ordinary conditions, the cell membrane prevents Na^+ from entering a cell by keeping the necessary channel closed.

On the other hand, as the channels for the abundant supply of K^+ in cells are always open, such ions possess a natural tendency to move outside of cells. Stated in a different way, the interior of cells becomes increasingly negatively charged as more and more of the positively charged K^+ move outside of the cells. Thus, from the electrical point of view, in order to compensate for the resulting imbalance, the positively charged K^+ are compelled to enter the cells.

Once equilibrium is achieved by this process, the flow of K^+ entering and leaving cells comes to a stop. At this point, the voltage within cells is referred to as the "resting membrane potential." This can be calculated by use of Nerunst's equation (3.1). The difference in electric potential, also known as the resting membrane potential (V), is therefore determined by the relative density inside and outside of the cells.

$$V \fallingdotseq -60 \log\left([K^+]_{in}/[K^+]_{out}\right) \qquad (3.1)$$

For this reason, the difference in the electric potential for all cells is approximately −90 mV towards the outside of the cell. However, considering the fact that Na^+ and Cl^- also gradually enter cells, the actual potential difference is about −70 mV.

However, in the case of the nerve cells, the membrane has a special function. For example, when the potential difference of −70 mV between the inside and outside of the cells decreases, a drastic change occurs. Specifically, once the membrane potential rises above the threshold, the normally closed Na^+ channels suddenly open to allow those ions to rush into the cells. Moreover, when membrane potential (V) suddenly increases in this way, an overshoot above 0 V occurs due to the inflow of positive ions (Fig. 3.2b).

If this change were to continue unimpeded, Na^+ would eventually fill up the cells to the point that they would collapse. However, while the Na^+ channels are closing, the K^+ channels are opening. Thus, a large quantity of the K^+ is released outside of the cells so that the membrane potential rapidly decreases, undershooting below the resting membrane potential.

The impulses generated in this way have a time frame of about 1 ms. By comparison, the period of undershoot lasts from 2 to 3 ms. The term "absolute refractory period" refers to the 1 ms period. Impulses are never generated during this period, regardless of the strength of the electrical stimulation being applied. On the other hand, the period between 2 and 3 ms is known as the "relative refractory period." During this period, an impulse occurs when the membrane potential is strongly raised, as the membrane potential is slightly lower than −70 mV. The Na^+ that once flowed into the cells must now leave, whereas K^+ reacts to the situation by entering the cells. At this point, a form of energy known as adenosine tri-phosphatase (ATP) functions as a pump to remove Na^+ while simultaneously drawing in K^+. The process is called the "Na^+-K^+ ion pump."

As Fig. 3.3 shows, when the impulses are conveyed to the synapses through nerve fibers, synaptic vesicles containing neurotransmitters move toward the synaptic gaps. Specifically, a synaptic gap between a synapse and the next neural cell has a narrow cleft, where synaptic vesicles emit neurotransmitters. If the neurotransmitters are excitatory in nature, the membrane potential of the next cell is raised, thereby generating an excitatory post synaptic potential (EPSP). Conversely, if the neurotransmitters are inhibitory in nature, the result is an inhibitory post-synaptic potential (IPSP). Impulses are generated whenever the spatial and temporal addition of the EPSP and IPSP goes beyond the threshold of the cells. Thus, the lateral inhibitory function works due to the fact that excitatory and inhibitory nerve cells combine in a lateral direction.

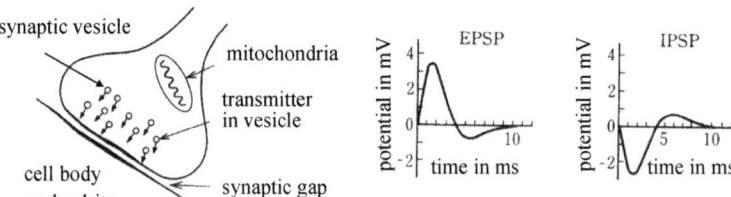

Fig. 3.3 Excitatory synapse and inhibitory synapse

3.2.2 Electrical Stimulation of the Auditory Nerves

When a negative pulse voltage is applied to certain nerve cell tissues through an adjacent electrode, the voltage difference between the inside and outside of the cells decreases. When this occurs, the ion channels of the membrane potential change, resulting in impulse generation. Actually, the medium from the electrode to the cells is not uniform: where the resistance is low, more electric current flows, while in the case of a high resistance, the flow of electric current decreases. Therefore, the membrane potential difference depends on the density of an electric current—not on the intensity of the added voltage. Furthermore, even from outside the cochlear canal, it is possible to fire nerve cells by applying a negative voltage to send an electric current to the inside of the cochlea. Based on this principle, various researchers have been striving to ascertain which method of electrical stimulation will prove most beneficial in recovering the auditory sense of the severely hearing impaired.

Among the hearing impaired, a malfunction in the hair cells is the main defect that characterizes those whose sensory neural hearing disabilities involving the inner ear. Although the function of hair cells deteriorates with age, other factors such as the use of antibiotics can also cause these cells to cease functioning. In other cases, a lack of hair cell function can be congenital. However, even in the case of non-functioning hair cells, if the nerves connected to the hair cells are still intact up to the CNS, it is still possible to convey sound information to the auditory field by stimulating the surviving nerves.

Full-scale research into cochlear implants began in earnest throughout the world after Simmons was able to clarify the main reason for pitch perception. His important finding was made possible from an experiment conducted in 1964, in which he inserted six electrodes into the auditory nerves of subjects in order to stimulate them.

His original idea was a rather simple one, based on the frequency pitch hypothesis. According to this principle, when electrical stimulation is simultaneously applied to nerves, the pitch changes according to the stimulation frequency. That is to say, when the nerves are stimulated by 100 Hz, a sensation of pitch corresponding to 100 Hz is perceived, while a stimulation of 200 Hz likewise results in a pitch sensation corresponding to that increased degree of stimulation. Since the range of a human voice pitch is at most from 80 to 300 Hz, it is possible to use such a method to convey pitch frequency. However, a frequency in excess of 500 Hz is not perceivable because the number of the nerve impulses is limited because of the absolute and relative refractory periods.

As described above, a very broad analysis of frequency occurs in the cochlea. More specifically, according to the place theory of the basilar membrane, the difference in sound frequency is converted into a difference in location inside the cochlea, and the perceived pitches are related to the location. For example, the entrance of the cochlea interprets high-frequency sounds, whereas the innermost portion of the cochlea deals with low-frequency sounds. Therefore, many researchers have adopted a method of conveying not only pitch but also spectral information based on the place theory. This method involves the insertion of an electrode array, consisting of a number of electrodes, into the cochlea.

3.2.3   Auditory Characteristics by Electrical Stimulation [3]

3.2.3.1   Single-Channel Stimulation

Any signal processing for cochlear implants should be designed in consideration of the sensory characteristics caused by electrical stimulation of the auditory nerves. Figure 3.4 schematically shows the relation between subjective values and a sine wave-like electrical stimulation of the auditory nerves [4]. First, the threshold and dynamic range that are barely perceived by the stimulation are indicated by the relation between the frequency and the intensity of the electrical current (Fig. 3.4a). The figure shows that the maximum loudness is less than 40 dB for a low stimulation frequency (around 150 Hz) while it is below 25 dB for a high frequency.

Next, the loudness estimation method was used to examine how loudness increases according to the increase in stimulation intensity. The results (Fig. 3.4b) show that loudness abruptly increases beyond B_S and that loudness is different between S_H at a high stimulation frequency and S_L at a low frequency. These characteristics are similar to those for the recruitment that is observed by the sensory neural hearing impaired. Therefore, although intonation of speech can be differentiated to some extent, information compression regarding loudness is essential in order to convey speech in the case of single-channel cochlear implants.

On the other hand, it is known that sensory temporal resolution is extremely high when the auditory nerves are electrically stimulated—even in the case of a single channel. For this reason, the author and his colleagues measured the discrimination

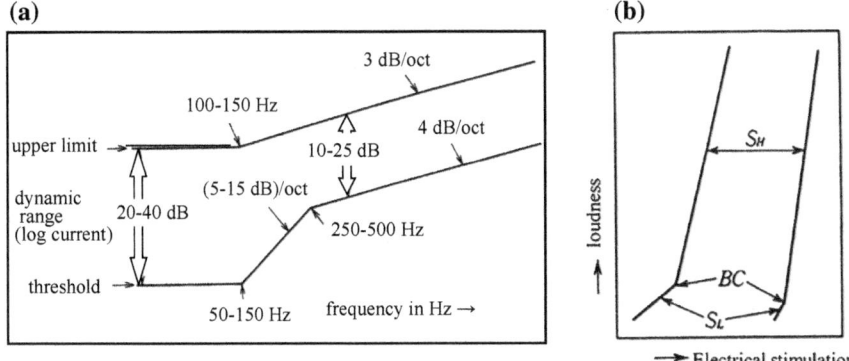

Fig. 3.4 a Dynamic ranges of sensation elicited by current stimulation to cochlear nerves. **b** Loudness by electrical current stimulation intensity

Fig. 3.5 Discrimination of
time difference Δt between
reference and test double
pulse stimulation

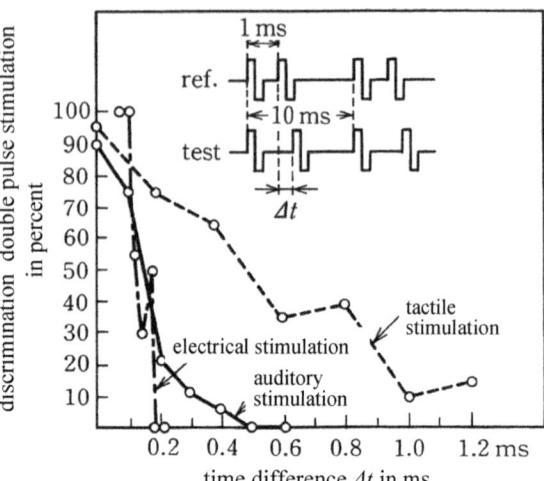

rate at which the reference stimulation ("ref." in Fig. 3.5) with a double pulse and test stimulation ("test" in Fig. 3.5) with another double pulse could be just detected when the pulse interval (Δt) for the test double stimulation was subtly changed [5]. In Fig. 3.5, the vertical axis indicates the discrimination rate and the horizontal axis indicates Δt. The dashed line shows the threshold difference for a time series of electrical stimulations consisting of two successive double pulses. From these findings, it is clear that the difference between the reference and the test stimulations cannot be discriminated with an increase of Δt of about 150 μs.

For the purpose of comparison, the same stimulation was administered to the auditory sense as sound. As in the case of electrical stimulation, the same tendency was observed (see solid line). However, from an experiment conducted to apply the same time series vibrations to the tactile sense of a fingertip, it was found that the difference threshold decreases by three or four times (see dotted line). It is anticipated that a method of converting phase information consisting of spectrum patterns into time series stimulation would prove effective for speech discrimination by electrical stimulation.

3.2.3.2 Multiple-Channel Stimulation

There are two factors in pitch perception that employ multiple-electrode methods. Figure 3.6 shows an example of perceived pitch versus electrical stimulation information obtained by an acquired deaf patient [6]. Three electrodes were inserted into the 8th nerve and the subject was asked to indicate the height (in inches) at which the pitch was perceived when the stimulating pulse frequency was changed. From this study, it became clear that the perceived pitch change depends on both the stimulation frequency and the place where the electrode is inserted.

Fig. 3.6 Perceived pitch as functions of stimulation frequency and stimulation electrode

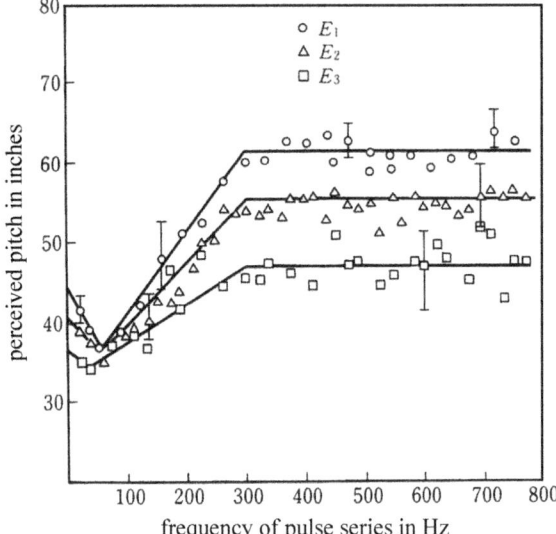

As Fig. 3.7 shows, Eddington conducted research to examine pitch sense by inserting six electrodes into a cochlea. By the use of any one of the electrodes, his findings revealed that as the stimulation frequency increased, the pitch also increased. In contrast, when the pitch frequency was fixed, the pitch sense depended on the location where the electrodes had been placed. Furthermore, although the pitch discrimination score increased when the pitch frequency was in the range of

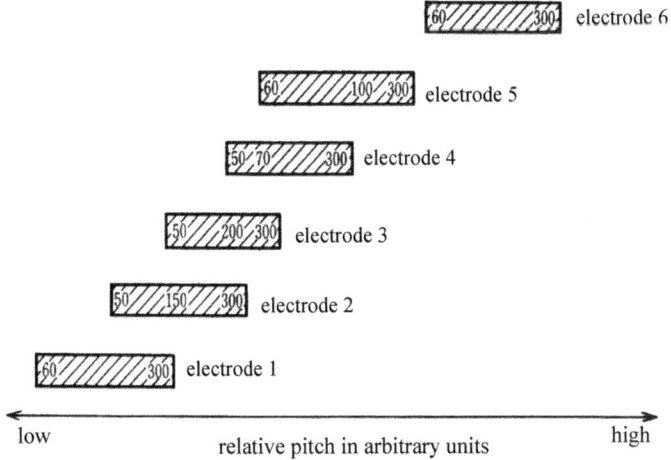

Fig. 3.7 Relative pitch elicited by stimulation point and stimulation frequency of a multichannel electrode array

80–500 Hz, once the frequency was increased to 400–500 Hz, the pitch discrimination score decreased notably [7].

The problem is that, even with the use of multiple electrodes, the electrical
current spreads out considerably, since the cochlea is filled with a conductive lymph
fluid that causes interference between the channels. In particular, the electrical
current pattern outside of the cochlea is extremely dispersed and complicated due to
the cochlea's spiral shape. Therefore, merely increasing the number of electrodes
does not necessarily mean that the amount of information conveyed likewise
increases. Thus, how to control this spreading of the electrical current is a major
problem with multiple-electrode methods.

3.2.4   Current Distribution Inside the Cochlea

Due to the fact that the lymph fluid inside the cochlea is an electrolyte solution, it is
difficult to limit electrical stimulation to specific nerves when such stimulation is
applied to a specific point inside the cochlea as the electrical current spreads out. As
Fig. 3.8 shows, the author and his co-researcher have also tried to measure the
distribution of electrical potential outside of the cochlea [8]. This was attempted by
first obtaining a part of a cochlea canal that had been removed from a cadaver and
affixing it to an acrylic cylinder. Next, an electrode array was inserted into the
cylinder in order to stimulate it with an electric current. As the left figure of Fig. 3.9
shows, our quantitative analysis revealed that the electrical current flows out at the
point where the auditory nerves leave the cochlea and that the distribution of the
electrical current is very widespread in that area. Furthermore, as the right figure of
Fig. 3.9b shows, when the electrical potential was measured outside of the cochlear
canal by inserting an electrode into it, it was clear that the electric current was
extremely widespread in addition to its being sent in the axial directions of the
cochlea.

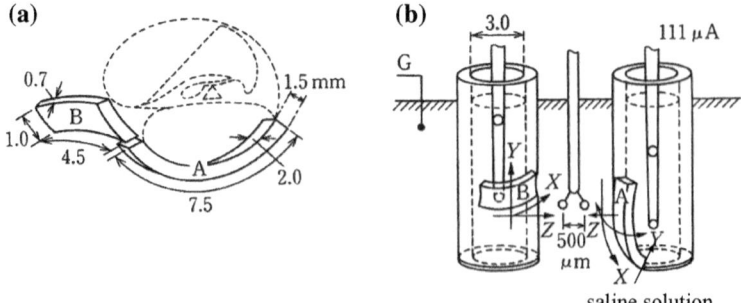

Fig. 3.8 Measurement method for potential distribution outside cochlea

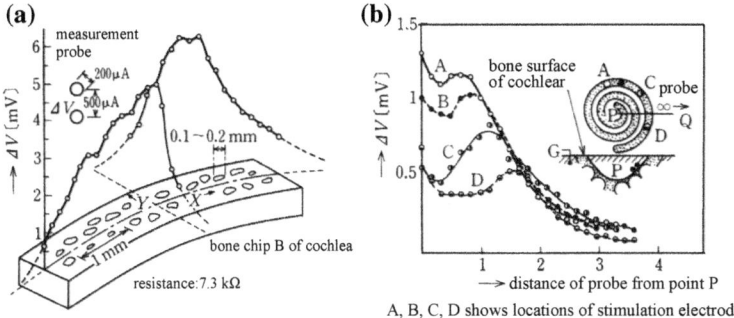

Fig. 3.9 *Left* Potential distribution flowing out from a bone chip of the cochlea. *Right* Potential distribution measured along the P-Q line, when stimulation electrodes were located at A, B, C, D

Therefore, accurate information about place pitch cannot be conveyed by simultaneously generating stimulations from two electrodes because they overlap each other. Practically speaking, it would be best if it were possible to connect the nerves and electrodes by inserting the electrodes directly into the nerves, as was strongly recommended by Simmons at the cochlear implant research group of Stanford University. However, this presents a major problem due to the fact that the nerves leaving the cochlea are spiral shaped, making it difficult to determine which nerves correspond to which frequency.

3.3 Various Approaches to Cochlear Implants

3.3.1 Electrode Configurations of Cochlear Implants

There are six different ideas regarding the electrode arrangement for cochlear implants. As Fig. 3.10 shows, they can be described as sliced cylinders, which indicate the cross sections of the cochlea. The multiple lines on the right side of the figure represent the auditory nerves. As shown in the figure, the electrode structures that have already been proposed can be basically classified into two types: the mono-electrode method and the multiple-electrode method. The former type can be further subdivided into the following three approaches: setting up the electrodes in the promontory or round window, located outside of the cochlea; placing the electrodes in the scala tympani inside the cochlea through the membrane of the round window (Fig. 3.10a); and inserting an electrode through a hole made in the cochlea (Fig. 3.10b).

In the case of the mono-electrode type, it is impossible to cause a person to perceive the height of a sound based on the place theory because many nerves are stimulated via the route to the opposite pole. Therefore, information must be

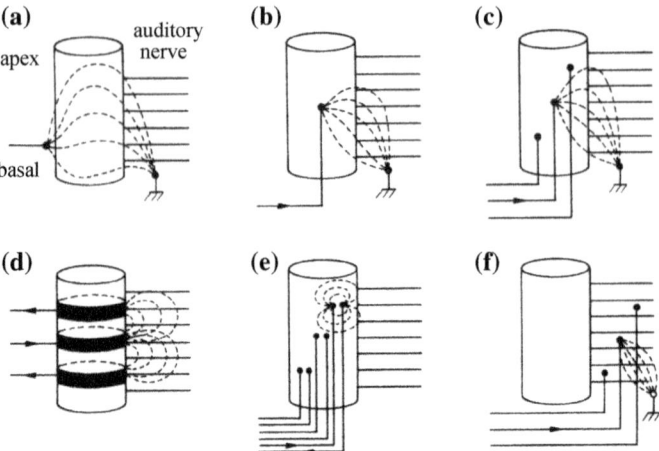

Fig. 3.10 Various arrangements of electrodes and stimulation methods

conveyed by frequency pitch, which makes a person perceive sound height according to how often the impulse fires, as mentioned earlier. Moreover, the volume of sound must be conveyed by the intensity of the current stimulation. When it is absolutely necessary to convey spectrum patterns, the spectrum must be converted into time series signals.

For the single channel cochlear implants, a Teflon-coated alloy composed of platinum (90%) and iridium (10%) is generally used for the electrodes. The reason for this composition is to ensure the durability and elasticity of the electrodes as well as to ensure the safety of living tissue. During this process, the indifferent electrode is inserted into the auditory bundle or placed on the bone outside of the cochlea. The use of this mono-electrode method, developed for practical use by House, was once quite widespread [9]. However, single-channel method could only enable a person to perceive a part of speech within a pitch range of 100–500 Hz and was limited to a 10 dB change in sound pressure. Consequently, the amount of information that could be conveyed through this approach was limited.

The multiple-electrode cochlear implants are based on the principle of the place theory. The mechanics involve stimulating nerves within a certain area by inserting multiple electrodes into the cochlea and generating electrical stimulations from specific electrodes according to the pitch of a sound. Therefore, in principle, the multiple-electrode method can be used to simultaneously convey not only frequency pitch but also spectrum information. Furthermore, in order to control the spread of the electrical current, the bipolar method (Fig. 3.10e) is effective. Specifically, this method involves fixing one side of the electrode as the positive pole and the other side as the negative pole. In this way, it is possible to stimulate nerves locally because the current flows in and out within a narrow range. In order to narrow the distribution of the electric current, most researchers of cochlear

Fig. 3.11 Schematic representation of 22-channel electrode array developed by research group at Melbourne University

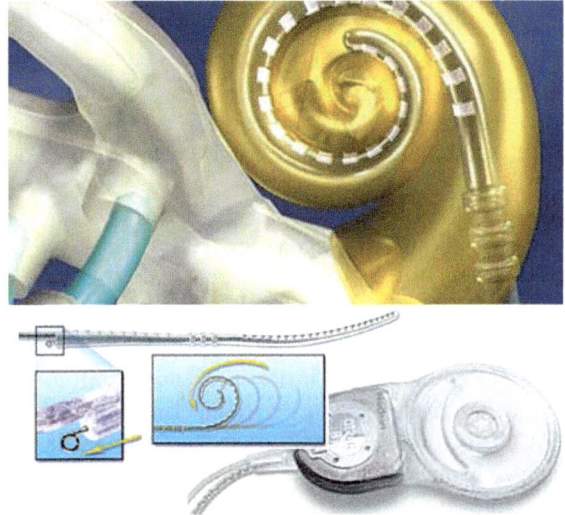

implants throughout the world switched from conventional mono electrodes to bipolar electrodes.

On the other hand, the method developed by Clark and his colleagues consists of using 22 platinum electrodes that are wound around a silicon holder 0.75 mm apart [10]. To prevent the distributions of electrical current from overlapping with each other, they are manipulated in such a way that only one of the 22 electrodes is activated at one time (Fig. 3.10d). The electrodes are inserted at a depth of around 24 mm into the scala tympani and are attached to the same location as in a normal cochlea, corresponding to the range of 600–8000 Hz. The electrodes are in the shape of rings, as shown in Fig. 3.11. This 22-channel interior cochlear implant was first produced by Nucleus, Ltd. in Australia (presently, Cochlear, Ltd. in the United States) and quickly became widespread throughout the world.

Unlike FES or a brain machine interface (BMI), artificial sensory systems involve afferent nerve stimulation, and much about where and how the transmitted information to the brain is later processed remains unknown. This makes it difficult to design artificial sensory organs. However, we do know that changes in brain function (plasticity) play an important role in the meaningful interpretation of new sensory stimuli. This offers promise in guiding the design of these devices. Yet much also remained unknown about brain plasticity. Thus, clinical application of cochlear implants was a trial-and-error process that required continuous evaluation and improvement at the beginning of the 1980s.

3.3.2   Signal Processing for Cochlear Implants (See [3])

Speech-recognition ability by cochlear implants exerts a profound influence on not only the electrode structure but also on the signal-processing methods. The latter consists of converting speech into the appropriate degree of electrical stimulation.

3.3.2.1   Single-Channel Method

In the case of a single channel, one signal-processing method is to convert speech waves through amplitude modulation. Another method involves stimulation by use of a pulse current that instantly changes from a negative to a positive charge by extracting only the pitch and intensity. In both cases, automatic gain control (AGC) is used to compensate for the small dynamic range. Although the single-channel methods are past studies, in order to offer a description of the basic principles involved in this method, the author will briefly mention an example of the mono-electrode type cochlear implants, developed by the House Ear Institute and manufactured by 3M Ltd. in the United States. As Fig. 3.12a shows, this method (House Urban type cochlear implant) consists of extracting the elements of 200–4000 Hz sounds from voice signals in order to create 16 kHz carrier wave amplitude modulation by means of a modulator. As Fig. 3.12b shows, the modulator is characterized by a high degree of nonlinearity between the microphone's output and the external coil's output. The degree of nonlinearity can be customized for each individual by adjusting the sound volume and modulator [11].

On the other hand, Hochmair made use of a more direct method, utilizing only the functions of intensity compression and frequency characteristic conversion [12]. This is in contrast to various other methods that analyze the characteristics of extracted voice sounds or convert voice sounds into pulse stimulations. It is said that since Hochmair's approach is simpler, the necessary equipment can also be more compact. Additionally, this approach facilitates rehabilitation by making it possible to perceive natural sounds. Researchers using this method believe that pulse stimulation is not very suitable for conveying voice sounds, especially since a method for encoding pulse stimulation remains unknown. Moreover, another disadvantage of using pulse stimulation is the difficulty in converting environmental sounds into natural sounds.

The author and his colleagues once developed a single-electrode type cochlear implant that initiated stimulation from outside the cochlea (Fig. 3.13) [13]. Our method involved first covering a Pt–Ir electrode with silicone and then affixing it to the round window. As this simple method sends signals through the outer ear, it has the advantage of even being applicable to infants since it is non-invasive stimulation.

Using this method, spectrum patterns were converted into time series stimulations, and a certain number of tones could be conveyed. For example, it was possible to convey the information about a second formant through two successive

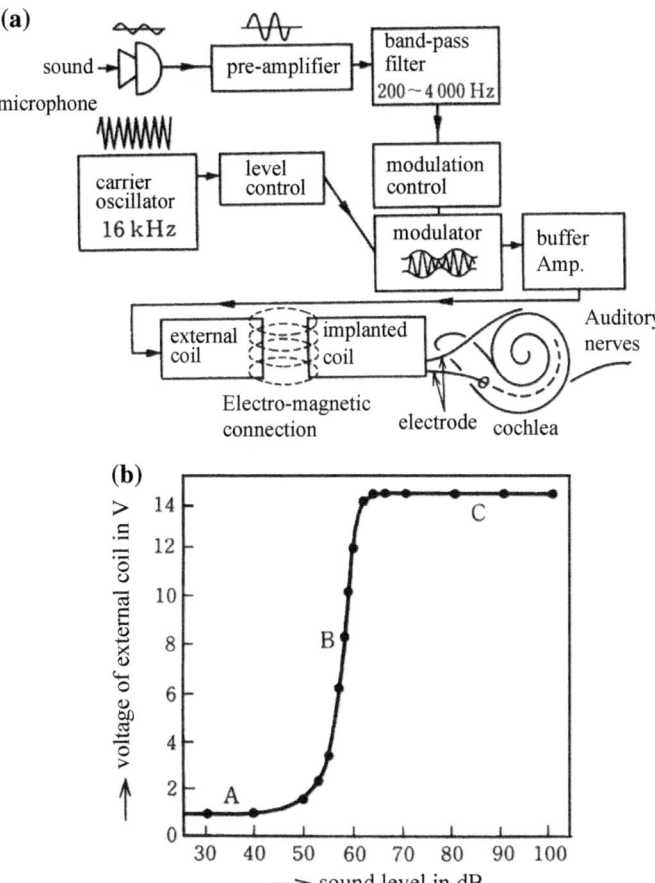

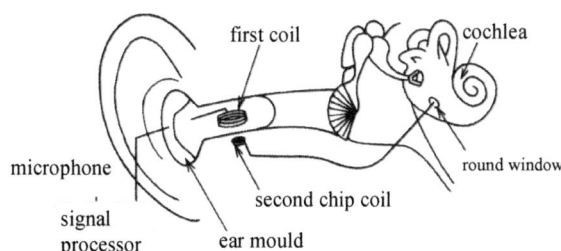

Fig. 3.12 a Block diagram of house urban's single-channel cochlear implant. **b** Input-output characteristics of house urban-type cochlear implant

Fig. 3.13 Configuration of single-channel stimulator for non-invasive auditory stimulation

pulses, known as double-pulse stimulations, that are generated in synchronization with the speech pitch, while the time interval (Δt) between two successive stimulation pulses was proportional to the reciprocal of the second formant frequency. When an experiment was conducted to distinguish five Japanese vowel sounds, the accuracy rate was found to be about 60%, although the perceived stimulations were very different from the sounds that the subjects actually remembered hearing.

3.3.2.2 Comparison of a Multiple-Channel Method Using a Single Channel

In the case of multiple-electrode methods, the basic process is to compress speech signals to break them up into frequency resolutions of certain channels so that an electric current corresponding to the proportional intensity of each element can be output from the corresponding electrodes. For example, in Eddington's method, the frequency range of speech signals is divided into four channel through a band-pass filter, each is compressed, and they are then converted into electrical stimulations to electrify each electrode (Fig. 3.14) [14]. Eddington showed that the method of using four channels with filters was much better than the methods of using four channels without filters and a single channel without filters for discrimination of a stimulus pair of synthetic speech sounds such as /da/-.ta/, /ba/-/ga/ as shown in Fig. 3.15.

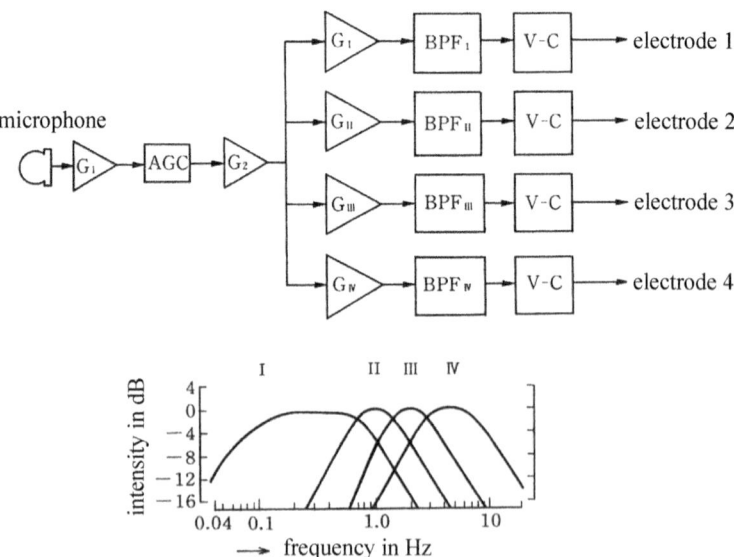

Fig. 3.14 Signal processing for 4-channel cochlear implant and characteristics of 4 band-pass filters

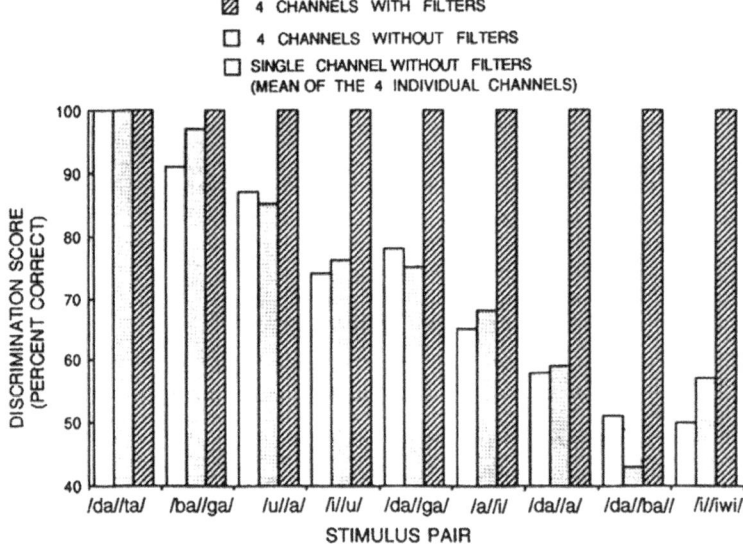

Fig. 3.15 Histogram of discrimination scores for stimulus pair of synthetic speech sounds

3.3.2.3 Current Sharpening Method

The author took part in joint research at Stanford University in 1984 and designed a cochlear implant with a sharpening function for stimulation current spread inside the cochlea [15]. As seen in the left figure of Fig. 3.16, instead of mono electrodes, research conducted at that time utilized 8-channel poly electrodes. Before implanting them into a human body, an 8-channel electrode was inserted into a cochlea removed from a dead monkey. This procedure was done to first verify the proper insertion of the electrode. Seen through a microscope, the top right figure of Fig. 3.16 shows a picture of a sliced cochlea with a black electrode and nerves spreading out in a radial pattern.

Figure 3.17 shows how the signal processing was accomplished. To control the distributional spreading of the electrical current in the cochlea, a lateral inhibition circuit was applied to the 8-channel function. At Stanford, Japanese speech recognition was examined on the basis of the Japanese phonetic syllables—/a/, /i/, /u/, /e/, /o/—after first implanting a signal-processing device into the cochlea of a human body and connecting it to an electrode array. In particular, the author and his co-researchers examined the effect that the lateral inhibitory circuit has over vowel recognition. Our findings indicated that the effect of lateral inhibition on the recognition of vowel sounds by synthesized vowels was unclear.

However, when the relationship between stimulation intensity and the perceived scale was checked, it became clear that lateral inhibition had a smaller inclination of loudness towards the stimulation level and allowed for the perception of sounds within a stronger dynamic range (Fig. 3.18). However, the perceived sounds were

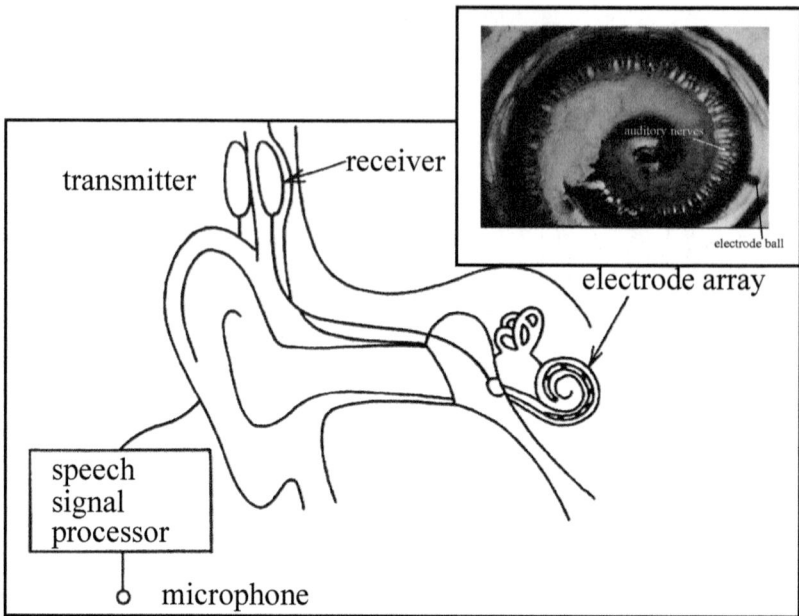

Fig. 3.16 *Left* Schematic representation of 8-channel cochlear implant. *Top right* Electrode array implanted into monkey's cochlea

very different from the speech that had been remembered. Although it was possible to distinguish sounds such as /a/, /i/, /u/, /e/, /o/, the listener answered when asked that the perceived stimulation was a mere metallic noise at the beginning of the stimulation. However, after two to three weeks of training, the listener said that the stimulation sounded like speech sounds. This supported the assumption that the reason the stimulation was heard as speech-like sounds was because plasticity occurred within the listener's brain.

It was also reported that when the alternating current outlet of the band-pass filter was directly converted into electrical stimulation, the sound was perceived to more closely resemble a human voice. Thus, regarding cochlear implants, the sound tone seems to vary greatly depending on whether the stimulation is converted into a pulse and whether the fired timing of the nerves occurs in the same phase.

To solve the problem of current distribution inside the cochlear, Miyoshi and the author proposed a new auditory nerve stimulation method that we called the "tripolar electrode stimulation method" for cochlear implants. Our method provided stimulation by using three adjacent electrodes selected from among the electrodes of the electrode array, as shown in the top figure of Fig. 3.19. The center electrode receives the current emitted from the electrodes on both sides. This approach made it possible to sharpen the current distribution and to shift the stimulating points by changing the rate of the emission of the electric current. They also conducted animal experiments using this method. On the basis of the results, it was ascertained that the tripolar method may succeed in narrowing the stimulation region and

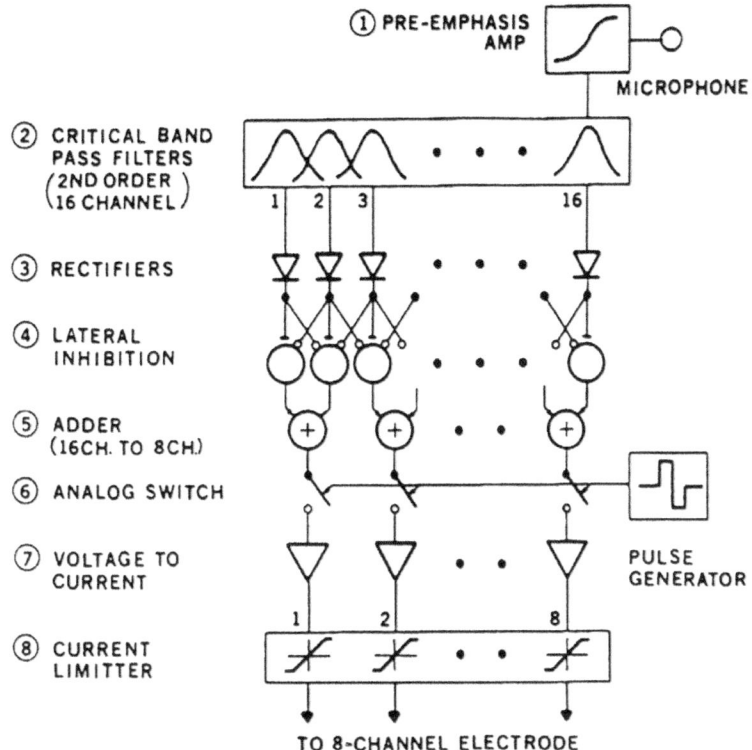

Fig. 3.17 Block diagram of speech processor for 8-channel implant with lateral inhibition

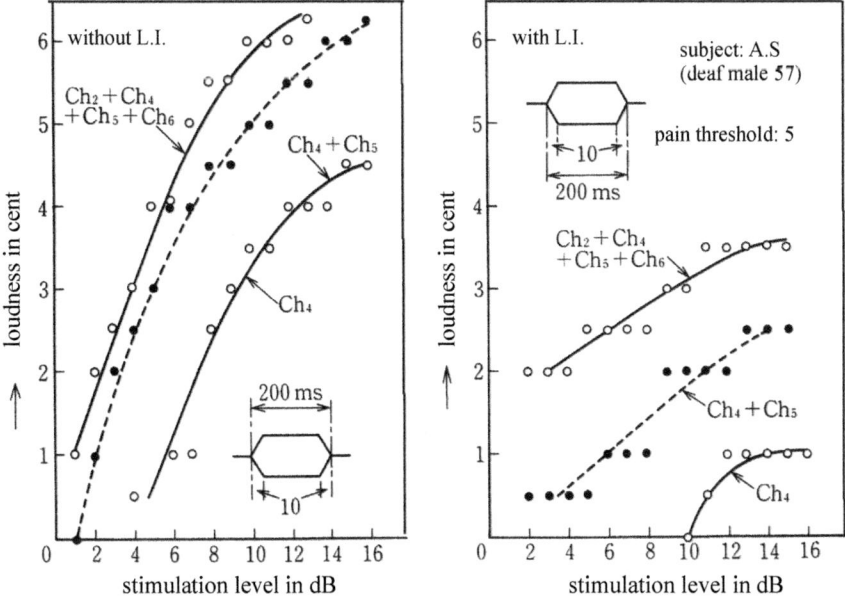

Fig. 3.18 Loudness as function stimulation level. *Left* Without lateral inhibition. *Right* With lateral inhibition

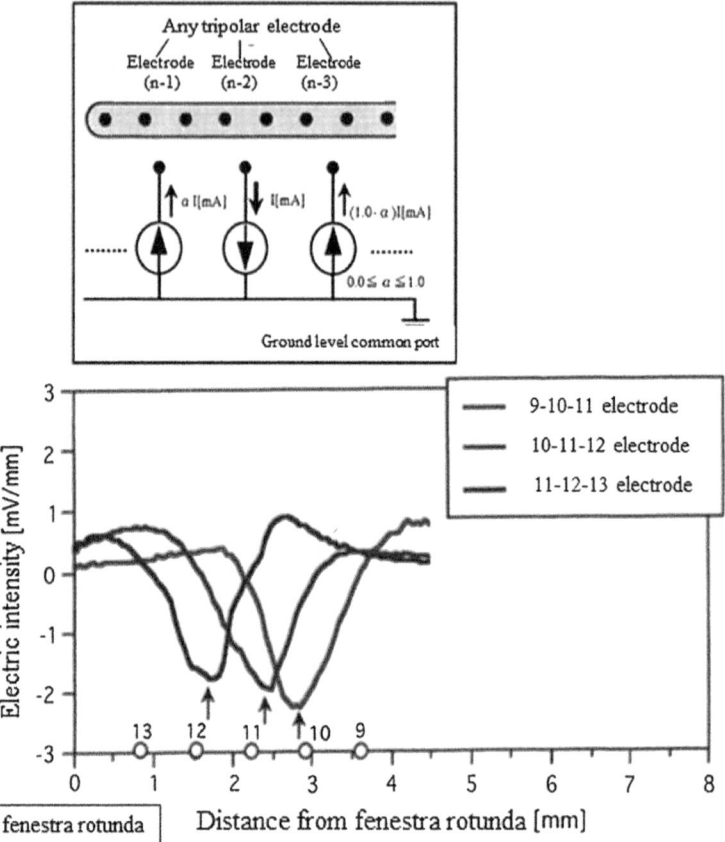

Fig. 3.19 *Top figure* Circuit architecture of tripolar-electrode stimulation method. *Bottom figure* Stimulation region and continuously moving stimulation site

continuously moving the stimulation site, as shown in the lower figure of Fig. 3.19 [16]. In contrast, the method used by most cochlear implants involves absorbing the electric current using only one electrode.

3.3.3 Configuration of 22-Channel Cochlear Implants

At present, the Melbourne method is the most widespread as a pioneering technique for practical use. As previously stated, this was developed by Clark and his colleagues and commercialized into a product by Nucleus, Ltd. (presently Cochlear, Ltd.). As mentioned above, this method uses 22 electrodes. It extracts the fundamental frequency (F_0), the first-formant frequency (F_1), the second-formant frequency (F_2) and so on, then selects electrodes that correspond to each frequency to

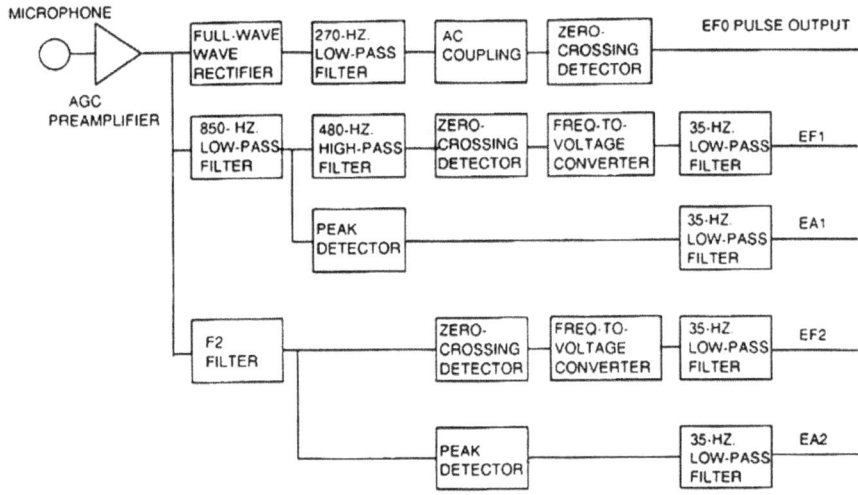

Fig. 3.20 Block diagram of wearable speech processor ($F_0F_1F_2$—type) of Nucleus 22-channel cochlear implant

electrify them. In order to prevent the electrical currents from overlapping each other inside the cochlea, this method controls the timing so that electrical current is not simultaneously generated from multiple electrodes.

Clark and his colleagues briefly reported that the fundamental frequency (F_0) combined with second-formant information (F_2) in this device is significantly better than the speech processor extracting the fundamental frequency (F_0) alone, based on results of both closed-set and open-set tests [17]. The $F_0F_1F_2$ speech coding strategy has been shown to result in improved understanding of speech compared with the (F_0–F_2) strategy. Significant improvement in closed-set speech recognition in the presence of background noise is an additional advantage of the $F_0F_1F_2$ strategy [18]. The $F_0F_1F_2$ speech coding strategy of the Nucleus 22-channel prosthesis is described in detail in a paper that Blamey and his colleagues [19] published in 1987.

As shown in Fig. 3.20, a hand-held microphone is used in the wearable speech processor and a 6-dB per octave pre-emphasis above 1000 Hz is switched into the circuit. An AGC method is used so that the maximum output corresponds to a comfortable level and also the stimulation is close to the threshold level during quiet periods. The maximum gain is variable in five steps: 32, 56, 66, 74, and 80 dB. The attack time of the AGC is about 1 ms and the decay time is between 25 and 50 ms. An amplitude compression ratio in excess of 25:1 is used.

First, the signal is full-wave rectified and filtered by a low-pass filter (LPF) to eliminate, as far as possible, the F_1 component. These envelopes with undulations are then passed through a zero-crossing detector so that F_0 is estimated. During unvoiced periods, the pitch extractor continues to examine the zero crossings of the signal envelope, but these excitations of the cochlea also occur at random intervals, producing a noise-like perception.

F_1 is estimated by low-pass filtering with a cut-off frequency of 850 Hz, two-pole filtering and then high-pass filtering by a single-pole filter at 480 Hz to remove the direct F_0 component. The filtered signal is passed through a zero-crossing detector, a frequency-to-voltage converter, and a 35-Hz low-pass filter to convert the zero-crossing rate linearly to level representing EF_1.

The F_2 component is estimated using a four-pole high-pass filter with poles placed at 850 and 1000 Hz so that, for an average of a large number of speakers, F_2 predominates over the other formants. After being shaped, the signal is passed through a zero-crossing detector and the resulting pulse train is fed into a frequency-to-voltage converter. The analog output of this converter is low-pass filtered to 35 Hz so that the smoothed EF_2 is able to vary with a response time of approximately 12 ms. As in the case of the F_0 estimation, the same circuitry is used for voiced and unvoiced sounds. Unvoiced sounds result in a high zero-crossing rate and produce stimulation on one of the more basal electrodes.

Each electrode pair is used to represent a frequency range of EF_1 and EF_2 in the speech coding scheme. The five most apical electrode pairs represent frequencies up to 1000 Hz. Their upper-frequency boundaries are equally spaced on a logarithmic scale from 300 to 1000 Hz and the remaining 15 electrode pairs are spaced on a logarithmic scale from 1000 to 4000 Hz. There is effectively an upward translation of frequency so that an EF_1 of 300 Hz produces stimulation in a region of the cochlea that would normally correspond to about 600 Hz, and an EF_2 of 4000 Hz would correspond to 8000 Hz in the cochlea.

The EF_0 is coded by the electrical pulse rate presented to the users. Whenever a zero crossing is detected by the F_0 estimation circuit, two electrical pulses are produced. The first pulse is applied to an electrode pair chosen according to EF_2, and the second pulse is applied to an electrode pair chosen on the basis of EF_1.

In addition, a diagnostic and programming system (DSP) is used for conducting psychophysical test on implant wearers and for configuring the user's wearable speech processor on the basis of these test results. Figure 3.21 shows the

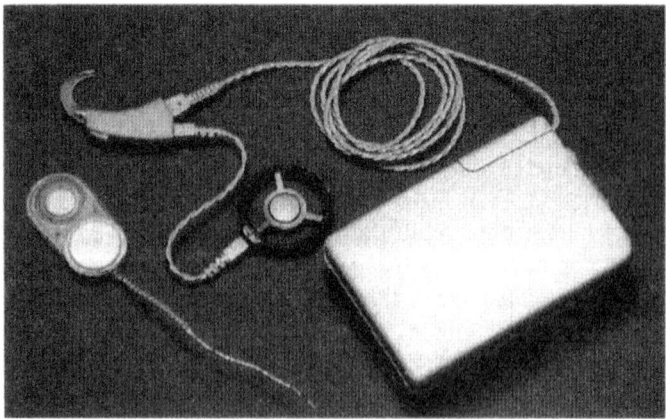

Fig. 3.21 Appearance of cochlear implant $F_0F_1F_2$ type produced by Cochlear LTD

Fig. 3.22 Electrical
stimulation pattern of
22-channel cochlear implant
for vocalization of /ada/

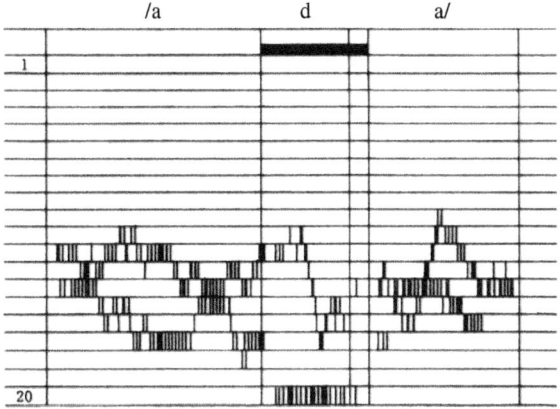

appearance of a cochlear implant made by Cochlear Ltd. Figure 3.22 shows the
electrode firing pattern following vocalization of the sound /ada/. In the consonant
portion of the electrode of Channel 20, firing corresponds to the pitch and is
characterized by a different pattern from /ata/. Methods of signal conveyance for
cochlear implants adopt an electromagnetic wave connection that consists of
making a hole in skull near the ear to implant a receiving circuit.

3.3.4   Speech Recognition by Cochlear Implants

It is assumed that there are around 16,000 hair cells that serve as auditory sense
receptors. Therefore, the quantity of information that can be conveyed through
electrodes is clearly limited since, at most, only 22 electrodes stimulate the hair
cells. Nonetheless, some users began perceiving electrical stimulation as sound after
having used cochlear implants for an extended period of time in the middle of the
1980s.

After Matsushima and his co-researchers conducted research under Clark, he
reported the first impression of a user who used cochlear implants. The subject was
a woman whose hearing had started to deteriorate at the age of 13 and who had
become completely deaf by middle age. Her impression was as follows [20]:

> Before the operation, I was told that I would hear some duck-like speech akin to that often
> heard in cartoons. But the first sounds I heard after switching on the cochlear implants was
> the chirping of cicadas. When I started using all 22 channels, I heard very high annoying
> sounds like whistles and pipes. In actual fact, what I was hearing was the voices of the
> people around me. Once I got used to the cochlear implants, however, human voices started
> sounding like metallic or throaty voices. At this point, I realized that the hearing aids were
> feeling more natural, since I was slowly getting used to them.

As I grew more and more accustomed to the cochlear implants, I found that I could recognize speech better than when I had used hearing aids. In fact, two weeks after I started using the cochlear implants, I was able to stop the use of hearing aids altogether. As training continued, I was able to recognize more and more sounds until I could eventually distinguish the sound of a telephone, dogs barking, birds chirping and even the sound of the rain. These sounds saved me from what had once been a lonely world, bereft of sound. Unfortunately, I still find that music is intolerable and I cannot yet distinguish human voices when they are mixed with noise.

Recently, cochlear implants are used even by infants and are somewhat effective. However, since the age of adaptability is considered to be one to two years or older, a clear diagnoses must be made after that time to determine if a patient's hearing is severely impaired or if they are totally deaf. It was also problematic that surviving cochlear nerves might be damaged once cochlear implants were inserted into the cochlea. In any case, as cochlear implants are not without defects, in addition to efforts at improving their ability to enhance the perception of speech, it is important to clarify the categories of impairment for which they can best be adapted. In spite of these arguments, 22-channel cochlear implants have been dramatically improved over the past 20 years.

The Clark-method model has quickly spread throughout the world with repeated model changes and is sold by Cochlear Ltd. in the United States. In Japan, Funasaka and his colleagues first adopted cochlear implants in clinical practice for the first time in 1987, and examined the resulting hearing ability for Japanese voice sounds. Kaga and his colleagues conducted many clinical tests to show the effectiveness of cochlear implants. As a result, coverage under health insurance become available for cochlear implants in 1994 in Japan. Recently, an average of approximately 500 persons among profoundly hearing-impaired patients are annually undergoing surgery for cochlear implantation and are regaining their hearing ability enough to engage in basic daily conversation, which is acclaimed as a successful example of artificial sensory organs.

3.4   Progress in Artificial Hearing [21]

3.4.1   Improvement in Cochlear Implants

The Clark-method model has been upgraded several times to the latest MULTIPEAK model, which has three band-pass filters (BPFs) in the upper-frequency region. Furthermore, the MULTIPEAK has been integrated into the SPEAK model that has 20 programmable BPFs. With both the MULTIPEAK and the SPEAK models, the output signals of the BPFs are scanned and then the peak outputs are transmitted into electric currents by scanning six out of 22 electrodes. Evaluation tests were conducted at Melbourne University Hospital using four cochlear implant models, and the results were compared from the viewpoint of speech-recognition ability [22].

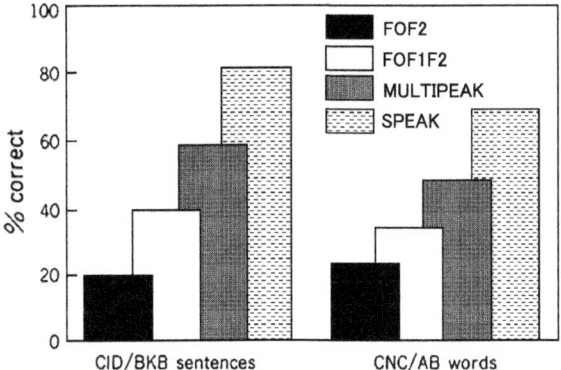

Fig. 3.23 Correct recognition rate of English sentences and words in daily conversation for 4 different models of the 22-channel cochlear implant

It became clear that the highest correct rate was obtained for the SPEAK model for both sentences and words used in daily conversation, as illustrated in Fig. 3.23 [23]. These results indicate that signal processing to extract formant frequencies is not necessarily needed and that, differing from the normal ear, subjects can hear voice sounds even when all auditory nerves are fired at once.

To date, cochlear implantation has enabled a lot of hearing-impaired people around the world to regain hearing to a level sufficient to carry on daily conversations. Thus cochlear implants are an example of successful artificial organs.

3.4.2 Other Methods of Cochlear Implants

In order to obtain natural firing patterns that are close to the patterns observed in normal cochlear nerves, Advanced Bionics Ltd. in the United States has produced a new cochlear implant model called the simultaneous analog sampling (SAS) method [24] that directly uses analog outputs of BPFs as stimulation currents (left figure of Fig. 3.24). As another approach, MED-EL Ltd. in Australia has designed a cochlear implant that uses the continuously interleaved sampling (CIS) method [25] in which stimulation pulses sweep over an electrode array (right figure of Fig. 3.24), the same as the vibration patterns of the basilar membrane.

In addition, basic research into a hybrid model has been formulated that places the cochlear implants only in a site corresponding to a high-pitched sound inside the cochlea, while storing the residual hearing in the low-pitched sound [26]. As a result of these improvements, if they have the cochlear implants installed during infancy, such stimulation experienced at an early age could enable users to develop a concept of language. Based on this reasoning, the adoptative age has recently been lowered to infants who are one year of age or even younger.

The results of investigations into the brain activity of people fitted with cochlear implants are providing many new findings about the recognition of spoken language. For example, the prefrontal area, which has been traditionally believed to be

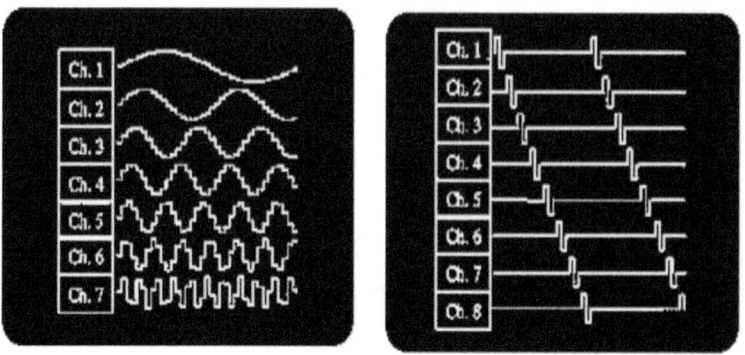

Fig. 3.24 *Left* Stimulation sequence for SAS method by Advanced Bionics Ltd., *Right* CIS method by MED-EL Ltd.

incapable of being activated by sound stimulation, can be activated immediately after the implantation of a cochlear implant.

Furthermore, some people who have cochlear implants fitted since an early age can enjoy listening to music, playing a piano or even a violin. These facts have led to new perspectives regarding research into language acquisition as well as in the brain function that works for music understanding. However, in a more general sense, the cochlear implants are not effective for all severe hearing-impaired people.

3.4.3   Auditory Brainstem Implants

One in 40,000 persons has neurofibromatosis, and the auditory nerve is often resected in those who undergo tumor resection for life-threatening complications. In these patients, a cochlear implant cannot be used, but auditory brainstem implants (ABIs), as shown in Fig. 3.25b, c, in which an electrode array is placed to stimulate remnant nerves to the CNS, may be promising [27]. The cochlear nucleus is systematically arranged according to sound frequency, so there is some correlation between electrode position and perceived pitch.

But in addition to the auditory nerve, the facial and glossopharyngeal nerves may also be stimulated. Thus, a maximum of only eight electrodes can be used. Newer devices with better frequency discrimination have been designed, but high surgical invasiveness and poor correlation with pitch are still a problem. Since 1979, several hundred patients have received ABIs worldwide. However, information transfer is limited compared to the cochlear implants, so that ABIs have been used together with lip reading.

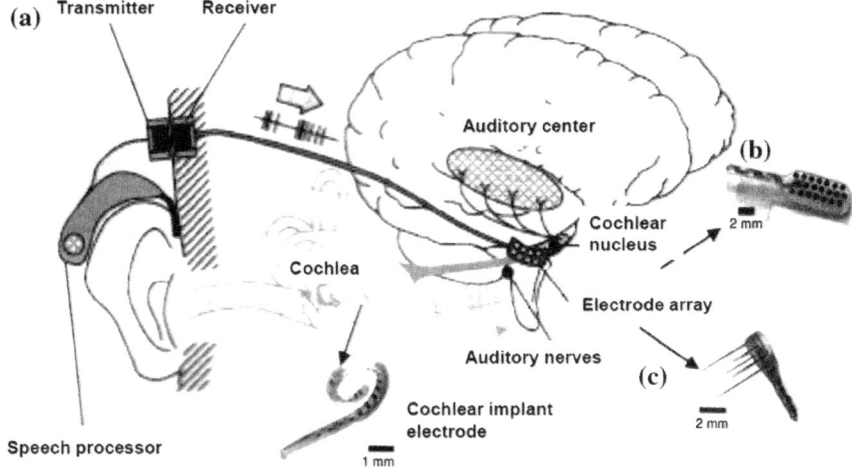

Fig. 3.25 Electrodes for artificial hearing and their implanted positions, **a** Cochlear implant, **b** ABI contacting type and **c** ABI inserting type

3.5 Secondary Effects Caused by Electrical Stimulation

While Matsushima and the author were developing a single-channel extra-cochlear stimulator, it became clear that electrical stimulation of the cochlea can lead to various side effects [28]. For example, tinnitus (ringing in the ears) can be cured, the head can feel refreshed, and hearing can be improved. Most interestingly, the surprisingly curative effect that this form of stimulation can have regarding the problem of tinnitus has been known for a long time. In fact, this phenomena also prompted us to conduct extensive research on this matter.

It is said that from 50 to 60% of the hearing impaired suffer from insomnia or stress caused by tinnitus, which is impossible for others to truly understand. Therefore, we developed a special implant to cure tinnitus that has a switch that the user can turn on once tinnitus starts. Although tinnitus retraining therapy (TRT) that relieves tinnitus by a kind of cognitive behavioral therapy has been in mainstream use, the author describes how this implanted device to cure tinnitus was developed and what its effects are [29].

Figure 3.26 is a circular graph representing the curative effect on 55 ears, including those of 24 men and 18 women. Their average age was 53.8 years. Tinnitus disappeared in five cases out of 55 and an improvement, including complete remission of the ringing sound, was observed in 70% of the subjects [30]. Furthermore, there was a tendency for the frequency of the tinnitus to switch from an annoying high-pitched tone to a less irritating lower tone. After applying electrical stimulation 152 times to the 55 ears, there were no examples of acute tympanitis, aggravation of defective hearing, or permanent tympanum perforation

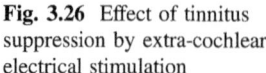

Fig. 3.26 Effect of tinnitus suppression by extra-cochlear electrical stimulation

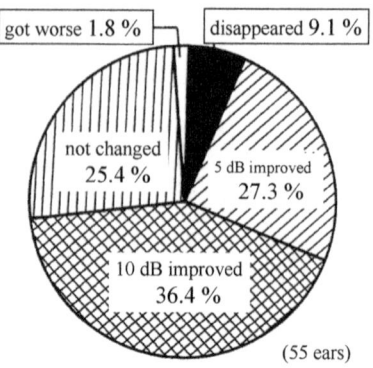

got worse 1.8 %   disappeared 9.1 %

not changed 25.4 %

5 dB improved 27.3 %

10 dB improved 36.4 %

(55 ears)

caused by tympanum incision. Thus, it is clear that electrical stimulation treatment for tinnitus constitutes an effective and safe method of treatment.

In almost half of the cases in which a curative effect was observed, electrical stimulation resulted in the temporary remission of tinnitus from a period of several hours to several days. Therefore, it would appear that tinnitus can be eliminated for an entire day if a patient undergoes, at most, several electrical stimulation treatments a day. If the tinnitus is the true cause of insomnia, it can also be stopped during sleep by applying electrical stimulation just before a person goes to bed and once after he/she has fallen asleep. The author and his associates have developed tinnitus curative devices that require a small invasive operation to provide stimulation from outside the cochlear.

The device consists of a coil in a plastic case for a hook-type hearing aid, plus another coil placed in the mastoid antrum combined with a stimulation device that shares the electric current (Fig. 3.27). In consideration of the user's convenience, the electric-stimulation device was attached in the same way as one wears a hook-type hearing aid. A silicone-coated coil was implanted inside the bone behind the ear while the patient was under general anesthesia. The stimulating pole of an electrode was placed in the middle ear and the opposite pole was placed in the tissue at the bottom of the implanted coil. The maximum quantity of electric current was set individually according to a limiter installed in the implanted tip coil.

To date, such devices to treat tinnitus have been applied to seven patients with positive results. Interestingly, the ringing sound heard in the opposite ear to the one being treated also subsided, leading to the assumption that the active part of the electric current that corresponds to the auditory system is actually the central nerve. In addition, after a few minutes of electrical-stimulation treatment, the tinnitus not only disappeared, but the users felt mentally refreshed and began to feel sleepy. With further stimulation, they fell asleep and reported that their refreshed feeling remained even after waking. The fact that the blood flow in their fingertips increased at this point indicated that the parasympathetic nervous system was dominant. Thus, it was thought that since electrical stimulation prior to bedtime

Fig. 3.27 Photograph of implantable tinnitus suppressor

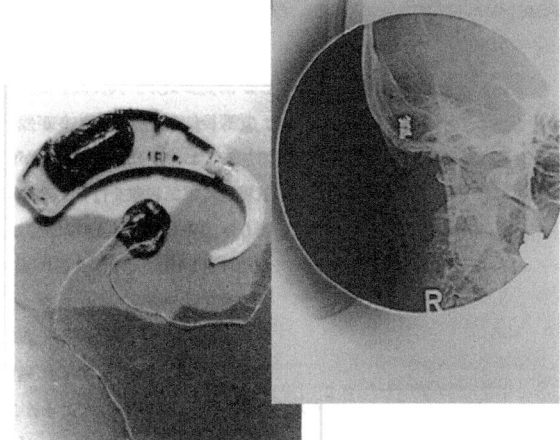

eliminated the ringing sound, they were able to sleep soundly, which, in turn, left them feeling refreshed following waking.

Differing from cochlear implants, this implant to treat tinnitus has no effect on sound perception but rather induces users to fall asleep because the parasympathetic nervous system becomes dominant. Furthermore, the refreshed feeling that users experience suggests that the limbic system, which plays a key role in emotion, is likewise effected. Although it is widely known that the musician Beethoven was hearing impaired, it is less well known that his loss of hearing nearly drove him mad. He must have experienced a great degree of stress due to his hearing loss, especially since the ability to hear is such a precious sense for musicians. It is worth noting that the above treatment is often unsuccessful for patients who suffer from some forms of mental illness. In particular, for those experiencing a high degree of mental stress, this treatment may prove ineffective regardless of the number of times it is applied. It is conceivable that the specific area where the tinnitus occurs is the CNS, even although it is the auditory nerves that receive direct electrical stimulation. It would appear that the main cause of tinnitus is stress, just as in the case of Beethoven.

3.6   Basic Concept and Future of Artificial Vision [31]

The success of cochlear implants has served as an impetus for research on artificial vision in many countries. In general, a device to electrically stimulate the optic nerve within the eye is called an "artificial retina." The broader term, which also encompasses direct stimulation of visual areas in the brain, is "artificial vision." In either case, the method involves division of the image detected by a camera into a grid pattern and stimulation with a 2-dimensional electrode matrix. However, how computer chips process the information that they provide to the retina and the significance of stimuli transmitted to the visual cortex require further investigation.

3.6.1 Artificial Vision

Research on artificial vision dates back to 1968 with a report published by Brindley et al. entitled "The sensations produced by electrical stimulation of the visual cortex" [32] (Fig. 3.28a). They reported "sensations of light (phosphenes) like stars at an unknown distance in the nighttime sky". Subsequent research on artificial vision was carried out by Dobelle. Over a 20-year period of implantable device use, 15-cm Landolt ring pattern recognition at a distance of 60 cm and no problems with infection were reported [33]. Despite feasibility issues and ethical concerns, this formed the basis of research in artificial vision.

3.6.2 Current Artificial Retinas

Artificial retinal implants are primarily indicated in partially blind patients with injury to the "macula-fovea" (where there is a high concentration of visual receptors) due to macular degeneration or retinitis pigmentosa. There are about 30 million such patients worldwide. Artificial retinas show the most promise in patients with residual bipolar cells (78%) and ganglion cells (30%), which are connected to visual receptors. In 1994, Dr. Liu developed a 4×4 electrode matrix for use in artificial retinas. These were implanted in volunteers with visual impairment.

About 10 years ago, the author attended an international symposium on artificial organs, spent 3 days with Dr. Liu, watched videos of the testing of artificial retinas, and participated in discussions on future directions in the field. According to Liu [34], when the electrode matrix (Fig. 3.28b) was first developed and implanted, patients were only able to detect whether a light or object was placed in front of them. But with improvements over a 10-year period, about 90% of recipients can now see faces and large figures, and some can even read up to 40 words per minute. However, to be able to read a newspaper, at least a 250×250 matrix is required. Even if this were to be developed, optimal placement in the retina and electrode wiring would be complicated issues that have not been resolved.

Artificial retinas have been used in several countries since 1995. In Japan, from 2001 to 2006, the Ministry of Economy, Trade, and Industry and the Ministry of Health, Labor, and Welfare sponsored a joint project entitled "Research and Development of an Artificial Vision System." The project was featured in this journal in 2006. The basic design uses suprachoroidal-transretinal stimulation (STS) (Fig. 3.28c). Instead of placement in the superior or inferior retina, electrodes are placed intrasclerally or suprachoroidally. This minimizes invasiveness to the

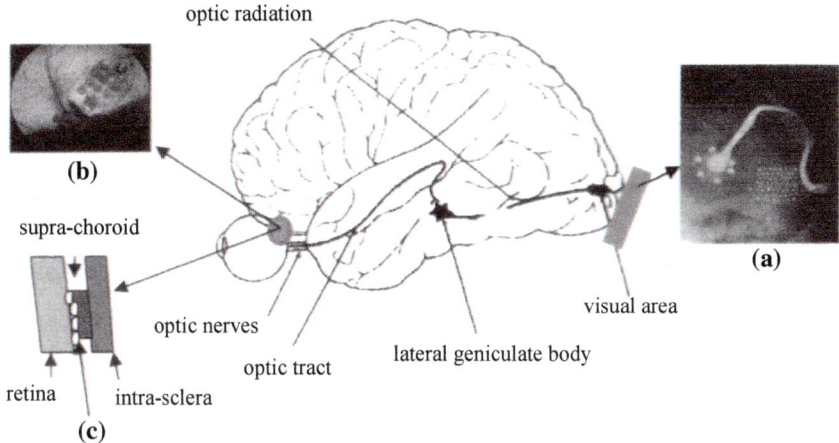

optic radiation

(b)

supra-choroid

optic nerves

retina   intra-sclera

(c)

optic tract   lateral geniculate body

visual area

(a)

Fig. 3.28 Electrodes for artificial vision/retinas and associated implant positions. **a** Artificial vision; **b** contact type of artificial retina; **c** STS (suprachoroidal-transretinal stimulation) type of artificial retina

retina before and after surgery. A trial device has already been evaluated in animal studies. The goal is to achieve visual acuity to "be able to count fingers at a distance of 30 cm". Clinical use of the device was expected by 2010 [35]. An example of a recently put into practical artificial retina is shown in Fig. 3.29.

Whether the brain's excellent plasticity in response to cochlear implant stimulation will be reproducible with respect to vision is still unanswered. Thus, at present, we cannot predict whether artificial retinal implants will enjoy the same success as cochlear implants.

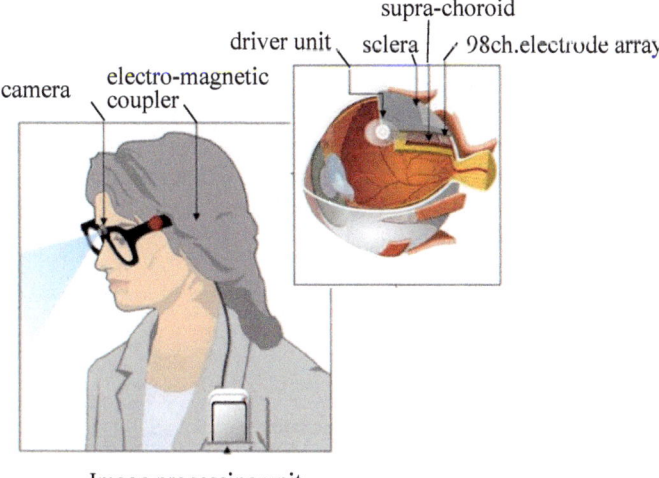

supra-choroid

driver unit   sclera   98ch.electrode array

camera   electro-magnetic coupler

Image processing unit

Fig. 3.29 An example of wearable artificial retina that stimulate visual nerves inside retina using 98 channel electrode array

3.7   The Future of Artificial Hearing

The first problem of cochlear implants is that an electric current spreads out inside the cochlea, as previously mentioned. Therefore, the quantity of information does not increase significantly merely because the number of electrodes has been increased. Actually, when the author compared the vowel sound discrimination score of an 8-channel cochlear implant with 8-channel tactile vocoder, as detailed in the next chapter, by the use of synthesized vowels, the tactile vocoder yielded almost the same results regarding vowel-recognition ability as cochlear implants.

The second obstacle is that electrode arrays can only be inserted halfway into the cochlea due to its spiral shape. As the arrays cannot reach the end of the cochlea, low sound information cannot be converted. Today, although the number of researchers in this area has decreased, those remaining researchers are tackling the twin problems of how to insert an electrode deeper into the cochlea as well as how to sharpen the distribution of the electric current. Hybrid type cochlear implants have been proposed as one method for transmitting low frequency sounds to the residual auditory nerve [36]. It goes without saying that even for the acquired hearing impaired, it is not an easy task to adapt themselves to the new stimulations because the nature of such stimulations could be different from what they remember if the signal process is imperfect.

The other problem concerns ethics and education. Some people worry about the inequality between those who experienced success with cochlear implants and those who did not, and also worry about the disappearance of the traditional way of acquiring language ability through special education or a finger language culture for the deaf. In the author's pursuit towards the final goal of improving overall communication for the hearing impaired, it has become clear that we must endeavor to find the most effective way of combining the functions of the auditory sense with those of the other senses [37].

At the same time, media conversion technology, such as speech-to-text conversion, is being used to help persons with visual and hearing impairment. With advances in regenerative medicine, information technology, education, and rehabilitation, the role of artificial sensory organs will continue to gain importance. Future knowledge of brain, visual, and hearing function may lead to cutting edge technology, that may be a "field of dreams" providing feedback on the development of even better artificial sensory organs.

References

1. A. Djourno, C. Eyries, Prothese auditive parexciation electrique a distance du nerf sensorial a l'aide d'un bobinage inclus a demeure. Press. Med. 14–17 (1957)
2. F.B. Simmons, Electrical stimulation for the auditory nerve in man. Arch. Otolarynol **84**, 2–5 (1966)

3. T. Ifukube (Eds. by Furui S, Sondhi MM), *Advances in Speech Signal Processing, Chapter 10, Signal Processing for Cochlear Implants* (Marcel Dekker Inc., New York, 1991), pp. 269–305

4. B.E. Pfinst, Operating ranges and intensity psychophysics for cochlear implants implication for speech processing strategies. Arch Otoiarygol, 110 (1984)

5. J. Matsushima, M. Kumagai, C. Harada, K. Takahashi, Y. Inuyama, T. Ifukube, A comparison of time resolution among auditory, tactile and promontory electrical stimulation—superiority of cochlear implants as human communication aids. J. Octolaryngol. Jpn. **95**, 1366–1371 (1992)

6. R.L. White, Review of current status of cochlear prostheses. IEEE Trans. BME **29**, 233–237 (1982)

7. D.K. Eddington, Speech discrimination in deaf subjects with cochlear implants. J. Acoust. Soc. Am. **68**, 885–891 (1980)

8. T. lfukube, R.L. White, Current distribution produced inside and outside the cochlea from a scala tympani electrode array. IEEE Trans. BME **34**(11), 876–882T (1987)

9. W.F. House, J. Urbun, Long term results of elelectrodeectrode implantation and stimulation of the cochlear in man. Ann. Otol. Rhinol. Laryngol. **82**, 504 (1973)

10. G.M. Clark, R.J. Hallworth, A multiple-electrode array for a cochlear implant. J. Laryngol. Otol. **90**, 623–627 (1976)

11. B.J. Edgerton et al., The effect of signal processing by the house-urban single-channel stimulator on auditory perception abilities of patients with cochlear implants. Ann. NY Acad. Sci. 311–322 (1983)

12. E.S. Hochmair, I.J. Hochmair-Desoyer, Percepts elicited by different speech-coding strategies. Ann. NY Acad. Sci. 268–279 (1983)

13. T. Ifukube, Y. Hirata, J. Matsushima, A new model of auditory prosthesis using a digital signal processor. J. Microcomput. Appl. Appl. **13**, 219–227 (1990)

14. D.K. Eddington, Speech recognition in deaf subjects with multichannel intracochlear electrodes. Ann. NY Acad. Sci. 241–258 (1983)

15. T. Ifukube, R.L. White, A speech processor with lateral inhibition for an eight channel cochlear implant and its evaluation. IEEE BME **34**, 876–882 (1987)

16. S. Miyoshi, S. Shimizu, J. Matsushima, T. Ifukube, Proposal of a new method for narrowing and moving the stimulated region of cochlear implants: animal experiment and numerical analysis. IEEE Trans. BME **46**(4), 451–460 (1999). doi:10.1109/10.752942

17. G.M. Clark, Y.C. Tong, R.C. Dowell, Comparison of two cochlear implant speech-processing strategies. Ann. Otol. Rhinol. Laryngol. **93**, 127–131 (1984)

18. B.K. Franz, R.C. Dowell, G.M. Clark, P.M. Seligman, J.E. Patrick, Recent development with the Nucleus 22 electrode cochlear implant: A new two formant speech coding strategy and its performance in background noise. Am. J. Otol. **8**, 516–518 (1987)

19. P.J. Blamey, G.M. CIark, Psycopysical studies relevant to the design of a digital electrotactile speech processor. J. Acoust. Soc. Am. **82**, 116–125 (1987)

20. J. Matsushima, M. Kumagai, K. Takahashi, N. Sakai, Y. Inuyama, Y. Sasaki, S. Miyoshi, T. Ifukube, A tinnitus case with an implanted electrical tinnitus suppressor. Nippon Jibiinkoka Gakkai Kaiho **97**(4), 661–667 (1994). [in Japanese]

21. T. Ifukube, Artificial organs: recent progress in artificial hearing and vision. J Artif Organs **12**, 8–10 (2009). doi:10.1007/s10047-008-0442-3

22. P. Blamey, P. Arndt, F. Bergeron, G. Bredberg, J. Brimacombe, G. Facer, J. Larky, B. Lindström, J. Nedzelski, A. Peterson, D. Shipp, S. Staller, L. Whitford, Factors affecting auditory performance of postlinguistically deaf adults using cochlear implants. Audiol Neuro. Otol. **1**, 293–306 (1996), doi:10.1159/000259212

23. T. Ifukube, Cochlear implants: progress of artificial organs. Japan. J. Rehabil. Med. **31**(4), 233–239 (1993). [in Japanese]

24. M.J. Osberger, L. Fisher, SAS-CIS preference study in postlingually deafened adults implanted with the CLARION cochlear implant. Ann. Otol. Rhinol. Laryngol **177**, 74–79 (1999)

25. J. Kiefer, J. Müller, Th Pfennigdorff, F. Schön et al., Speech understanding in quiet and in noise with the CIS speech coding strategy (MED-EL Combi-40) compared to the multipeak and spectral peak strategies (nucleus. ORL **58**, 127–135 (1996). doi:10.1159/000276812
26. B.J. Gantz, C. Turner, K.E. Gfeller, M.W. Lowder, Preservation of hearing in cochlear implant surgery: advantages of combined electrical and acoustical speech processing. Laryngoscope **115**(5), 796–802 (2005). doi:10.1097/01.MLG.0000157695.07536.D2
27. T. Ifukube, Artificial organs—recent progress "Artificial Hearing and Vision". J. Soc. Artif. Organs **12**, 8–10 (2009)
28. J. Matsushima, T. Kamada, N. Sakai, S. Miyoshi, M. Sakajiri, N. Uemi, T. Ifukube, Increased parasympathetic nerve tone in tinnitus patients following electrical promontory stimulation. Int. Tinnitus J. **2**(1), 67–71 (1996)
29. P.J. Jastreboff, Tinnitus retraining therapy. Textbook of Tinnitus, 575–596. doi. 10.1007/978-1-60761-145-5_73. Springer Science+Business Media, LLC 2011 (2011)
30. J. Matsushima, N. Sakai, S. Miyoshi, M. Sakajiri, N. Uemi, T. Ifukube, An experience of the usage of electrical tinnitus suppresser. Artif. Organs. **20**(8), 955–958 (1996)
31. T. Ifukube, *Artificial vision/retina* (Edit by JSAO, Haru Publishing Company, Artificial Organs Illustrated, 2012), pp. 395–417
32. G.S. Brindley, W.S. Lewin, The sensations produced by electrical stimulation of the visual cortex. J. Physiol. **196**, 479–493 (1968)
33. W.H. Dobelle, Artificial vision for the blind by connecting a television camera to the visual cortex. ASAIOJ **46**, 3–9 (2000)
34. W. Liu (2003) Intraocular retinal prosthesis: a decade of learning. Proc The 2003 International Workshop on Nano Bioelectronics, Seoul, Korea
35. Y. Terasawa, H. Tashiro, A. Uehara, T. Saitoh, M. Ozawa, T. Tokuda, J. Ohta, The development of a multichannel electrode array for retinal prostheses. J. J. Artif. Organs **9**, 263–266 (2006)
36. E.A. Woodson, L.A.J. Reiss, C.W. Turner, K. Gfeller, B.J. Gantz, The Hybrid cochlear implant: a review, in *Cochlear Implants and Hearing Preservation*, ed. by P. Heyning, P.A. Kleine **67**, 125–134 (2010). doi:10.1159/000262604
37. T. Ifukube, Discrimination of synthetic vowels by using tactile vocoder and a comparison that of an eight-channel cochlear Implant. IEEE Trans., BME **36**(11), 1085–1091 (1989)

Chapter 4
Tactile Stimulation Methods for the Deaf and/or Blind

Abstract In this chapter, the author introduces tactile stimulation methods for sensory substitutes of the auditory and/or the visual sense. Recent reports concerning brain research show that lost sensory functions might be compensated for by the other sensory cortex due to the plasticity of the neural network in the human brain. Based on the latest findings regarding brain plasticity, the author talks about the effectiveness of tactile stimulation methods that support speech recognition of the deaf, vocal training of the deaf-blind, and auditory localization substitutes. The author also describes how the findings and technologies obtained from tactile stimulation studies will lead to new concepts in the design of tactile displays used for virtual reality systems and robots.

4.1 A Role of Tactile Stimulation in Sensory Substitutes

Tactile aid is an auditory substitute method that transmits sounds by converting them into skin stimulation. Although research into tactile aids has a long history, it is not popular as a hearing substitute. However, it is premature to conclude that an adequate evaluation of tactile aids has been achieved as some researchers have drawn negative conclusions, while others have been more positive. The biggest unknown is whether it is possible to form word concepts through skin stimulation. Thus, the essential nature of tactile aids remains unknown. Furthermore, given the progress in cochlear implants, the number of people researching tactile aids worldwide decreased sharply for a while. Recently, however, researchers have found new applications for tactile devices such as for robots and virtual reality systems. Needless to say, since the tactile sense is the indispensable sense for the deaf-blind to communicate with people and also to acquire environment information the author shall attempt to describe it.

Many hearing-substitute methods appeal to the visual sense, as mentioned in Chap. 5. However, it is preferable to avoid the exclusive use of the visual sense as a

© Springer International Publishing AG 2017
T. Ifukube, *Sound-Based Assistive Technology*,
DOI 10.1007/978-3-319-47997-2_4

hearing substitute because occupying the visual sense may interfere with other important perceptual information. Since the tactile sense, in contrast, is spread throughout the body, its partial use as a hearing substitute will not interfere in a crucial way with overall perception. Moreover, both the tactile and auditory senses share the common characteristic of passively receiving information. Actually, in the past, several attempts were made to design a so-called "tactile vocoder" by means of vibrating 5–10 fingers to provide an alternate method of transmitting speech information.

When the tactile sense is used as a hearing substitute, it is necessary to explain the method for evaluating the receiving capacity of information to the tactile sense as well as the difference in the characteristic of mechanisms between the auditory and tactile senses. As the pitch and intensity of vocal sounds constantly change, the temporal characteristics of tactile sense information processing should be accurately understood. Moreover, when the amount of information in vocal sounds cannot be completely contained within the tactile information capacity, it is necessary to consider a method for compressing the information in the sounds. It is also necessary to consider the question of whether the human cortex has the capability of perfect association of tactile speech cognition to language conception.

The concept of tactile aids originates from Braille for the blind. Most blind people who learnt Braille can read about 10 Braille characters per second with their fingertips. In the case of speech, about five monosyllabic sounds per second are pronounced. For example if you say /i, ro, ha, ni, ho/ (the Japanese equivalent of "a, b, c ...") at a normal speed, it takes about one second. Therefore, when speech sounds are converted into an alternate information code, it must be possible to easily recognize five monosyllabic sounds per second with a fingertip.

According to neuroscience research conducted in 1998, it was reported that finger Braille patterns could reach the visual cortex through the tactile cortex and then the patterns would be associated with the language-understanding cortex, as shown in the left part of Fig. 4.1 [1]. Based on the plasticity of the neural network in the human brain, lost functions might be compensated for by another sensory cortex. In fact, in 1998, Swedish brain researchers used a functional MRI to show that vibro-tactile stimulation made from vocal sounds activates the auditory cortex of the acquired deaf as shown in the right part of Fig. 4.1 [2]. If tactile aids are worn from infant, it might be possible to understand languages.

In this chapter, the author will consider the capabilities of tactile aids by briefly referring to the physiological properties of mechanical receptors inside a finger and by describing the similarities of the tactile and auditory senses from the viewpoint of the psychophysical characteristics of both senses.

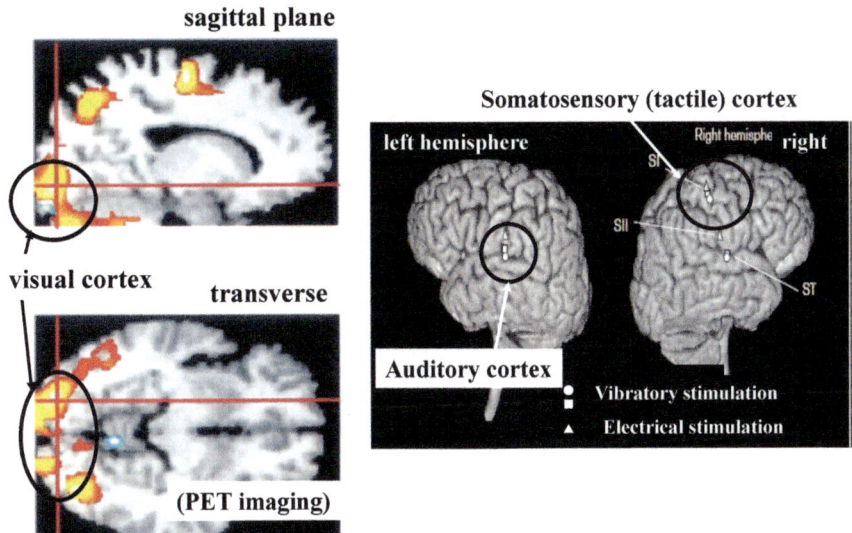

Fig. 4.1 *Left* Visual cortex (PET imaging) activated by Braille stimulation on the fingertips of acquired blind person. *Right* Auditory cortex (MRI of *left hemisphere*) of acquired deaf person activated by tactile stimulation

4.2   Basic Characteristics of the Tactile Sense

4.2.1   *Mechanical Receptors*

Psychologists have divided mechanical receptors into three main categories. The categories and their corresponding definitions are as follows: when the sense is characterized by the skin's deformation lasting for a certain period of time, it is called the "pressure sense"; when the skin is deformed only once but returns to its normal state quickly, it is called the "touch sense"; and when the skin is deformed continuously for more than 10 times a second, it is called the "vibratory sense". From the physiological point of view, a great deal of research has clarified that the receptors corresponding to each sense exist in the skin, particularly in the fingertips.

A cross section of the hairless part of a fingertip as seen through a microscope clearly shows the skin's "epidermis" and "dermis" layers (Fig. 4.2). The epidermis is about 1 mm thick and is very hard, while the dermis is 2–3 mm thick and is very soft. Fingerprints on the surface of the epidermis play an important role. For example, if the skin were smooth without any fingerprints and was pricked by a sharp object such as a needle, the tension of the skin would greatly expand. However, the advantage of having fingerprints is that only the area pricked by the sharp object is deformed so that the tension does not expand. Consequently, this characteristic improves spatial resolution in such a way that even subtle bumps temporarily change their shape. Accordingly, bumps of even 5–6 μm can be detected.

Fig. 4.2 Four tactile receptors (SA I, RA I, SA II, RA II) receiving mechanical stimulation and their characteristics

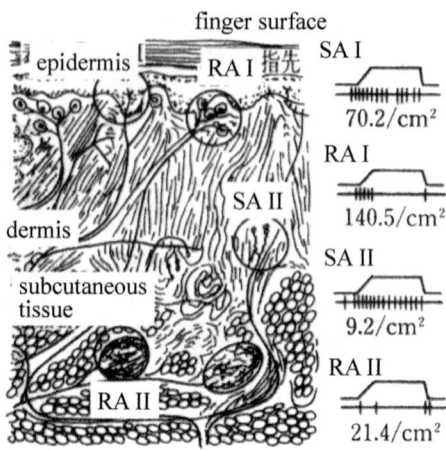

The fact that the epidermal and dermal layers differ in their hardness indicates that their mechanical impedance also differs. This area of the skin contains mechanical receptors. Impulses are generated because the energy difference caused by the change in mechanical impedance is transformed into receptor potential.

Meissner corpuscles (RA I in Fig. 4.2) are located on the border between the epidermis and dermis. If the pressure applied to the receptors of the fingers has a time pattern that forms a trapezoid starting from 0, increasing to a plateau and finally returning to 0, the receptor potential of Meissner corpuscles causes the generation of impulses at the parts where the pressure changes. Impulse patterns and the number of receptors per square centimeter are shown in the right side of Fig. 4.2. It can be said that this is a sensory response to linear differential, which is most sensitive when the vibratory frequency of the surface is around 40 Hz.

Furthermore, similar to Meissner corpuscles, there are also receptors called Merkel cells (SA I) between the epidermis and dermis. Johnson and his colleagues performed an experiment by applying Braille-like patterns to the fingertip of a monkey, and from the responding impulse pattern it was presumed that mainly Merkel cells are responsible for Braille perception [3].

There are receptors, known as Ruffini endings (SA II) that can be found at a slightly deeper level within the skin. Judging from the impulse patterns, it can be said that they also respond to pressure. Specifically, it is believed that they detect deformations in a horizontal direction, such as horizontal pressure to the skin of a finger. As vocal sounds can change at any moment, it is assumed that SA I and SA II have little to do with the conversion of vocal sounds into mechanical stimulation.

At a deeper level within the skin, there are receptors called Pacinian corpuscles (RA II). Each of them is onion shaped and reaches a size as large as 600 μm. It is known that they generate impulses only when pressure begins to change or stop. This occurs when pressure is applied to them, as in the example mentioned above. It can be said that they are receptors whose function is to respond to secondary differentials and to easily respond to vibratory stimulation with a high frequency. By examining how impulses are generated, it can be seen that they peak when the frequency reaches about 200 Hz (see, RA II in Fig. 4.2). In other words, sensitivity is best at around 200 Hz. Given that Pacinian corpuscles respond quickly, by effectively stimulating them, it is possible to convey a part of vocal sounds [4].

However, careful attention should be paid to the fact that the threshold fluctuates greatly, depending on the skin temperature. Moreover, the threshold increases over time if the duration of the stimulation is long because adaptation occurs quickly. Finally, the spatial resolution would be extremely low in this case because the number of receptors is small and they are deep within the skin. Actually, it is assumed that a low spatial resolution would be compensated for by a sort of sharpening function of the lateral inhibition neural network.

Incidentally, as shown in Fig. 4.3, the threshold of each receptor versus vibratory stimulation frequencies are different. Furthermore, although the mechanical characteristic of the four receptors mentioned the above are measured in the fingertip, in fact, the distribution density of the four receptors are different as shown in Fig. 4.4b. Therefore, the spatial resolution also greatly changes depending on the skin surface. In fact, the spatial resolution is highest in the fingertip, whereas, it is lowest in the palm as shown in Fig. 4.4a. In addition, transmitted information will be different between the "passive" way that a hand and fingers are fixed and "haptic" way that the hand or fingers are moving to receive the mechanical stimulation.

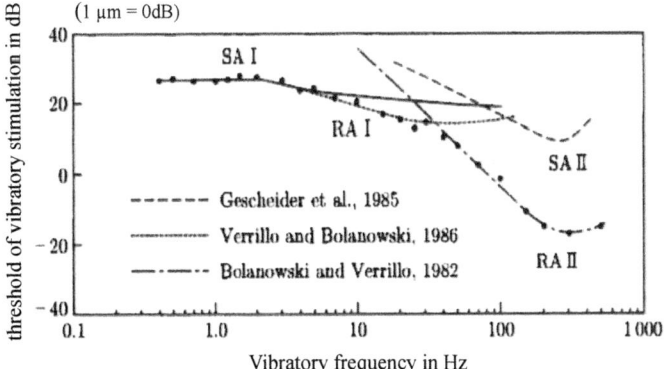

Fig. 4.3 Threshold-vibratory frequency characteristics of four tactile sensors inside a fingertip of human. The figure shows the data that were measured by three researchers

Fig. 4.4 a Dependence of
spatial resolution.
b Distribution densities of the
four mechanical sensors

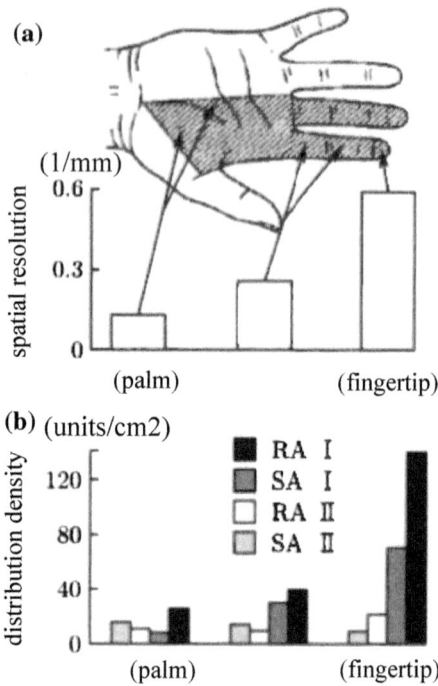

4.3 Spatial and Temporal Characteristics
of the Tactile Sense

4.3.1 Representation of Spatial Characteristics
Based on Psychophysics

As mentioned in Chap. 1, when two vibratory stimuli were added simultaneously to
a fingertip surface, the maximum transmitted information changed depending on the
distance between two stimuli and the stimulation. The result means that the subjective intensity (S′) and the subjective number (n′) of the multiple stimulation
change due to the suppressive effect and the overlap of the receptive field that may
be explained by using the neural unit proposed by Békésy [5]. The author proposed
how to estimate the neural unit as well as the subjective intensity (S′) and the
subjective number (n′) using the spatial masking value and the two-point threshold
[see 1.19].

In the experiment to obtain the two-point threshold, the skin surface of the
fingertip was first placed on a vibrator array, followed by the randomly chosen
vibratory stimulation (200 Hz for 0.5 s) of 2 out of 16 possible stimulation points,
with an interval of 2–3 s between each two-point stimulation. After each stimulation, the subjects were asked to indicate whether or not the stimulation was felt at

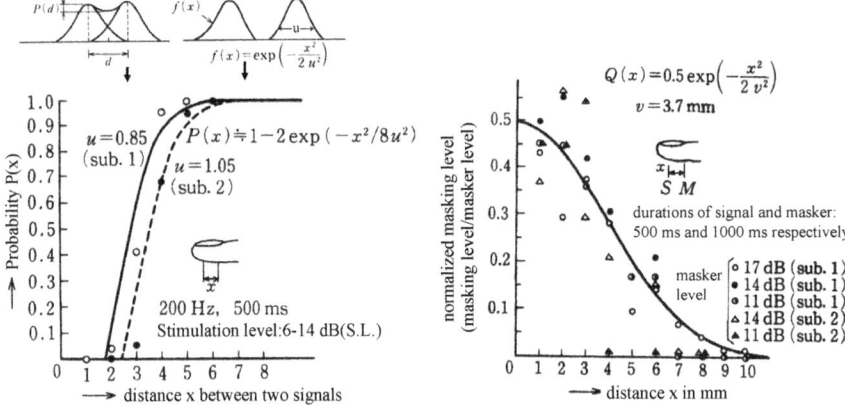

Fig. 4.5 *Left* Probability that two-point stimulations were separately perceived as function of distance between two points. *Right* Normalized masking level as function of distance between masker and signal in the tactile sense

one or two points. From the experimental result, it was ascertained that the relation between the probability $P(x)$ of a subject's sensation of two points and the actual distance x [mm] between those two stimulation points is as shown in Fig. 4.5.

From the figure, as x decreased, so also did $P(x)$, so that when $P(x) = 0.5$, the value of x is 2.5–3.5 mm. It goes without saying that two stimulation points were perceived as one because the subjects actually sensed only one overlapping stimulatory point. In other words, the transmission of the two separate vibrations was spread over a single point of their receptive fields. When the author approximated this spreading of the tactile sense by a normal distribution density function $f(x) = \exp(-x^2/u^2)$, the crest and trough of each wave appeared as a result of the spatial overlap of the two points of stimulation (see the upper figure of Fig. 4.5). Therefore, if $P(d)$ represents the level difference between the crest and trough for a distance d between the two stimuli, $P(d)$ is estimated by the equation shown in the left of Fig. 4.5. In fact, by using the least squares method, it was possible to obtain an approximation of the $P(d)$ with relative accuracy, as the solid line in the figure indicates. Furthermore, the value u that corresponds the spreading of the receptive field was also estimated to be about 2.0 mm.

Next, based on the masking experiment, it will be explained how the signal threshold changes, depending on the distance x between two stimulation points when the masker and the signal exist at the same time. In this experiment, the threshold (0 dB) for the signal stimulation was first determined. Then, the increase in the signal threshold ($S[dB]$) was measured when the masker was present. That amount was determined to be the masking level (M).

As a result, it was found that the masking level for the signals gradually increases as x decreases, and that the area affected by the masking is about 7 mm as

Fig. 4.6 Schematic
representation of excitatory
area and inhibitory area

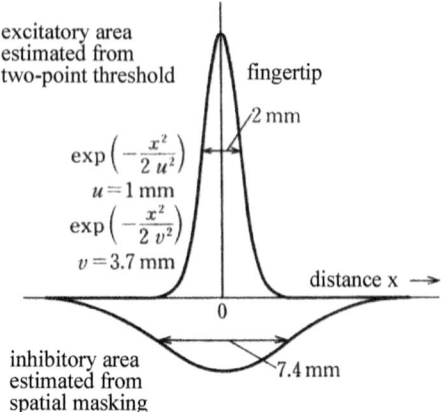

shown in the right of Fig. 4.5. Moreover, the vertical axis represents the masking coefficient, which equals the ratio of the masking level to the intensity of the masker. The solid line $Q(x)$ indicates the normalized masking level by the masker level.

$$Q(x) = a \cdot \exp\left(-x^2/2v^2\right)$$

where, a = 0.5, v = 3.7 mm

As $Q(x)$ indicates the scale and expanse of the suppression of the signal by the masker, the subjective intensity of the signal S' is estimated from $Q(x)$.

Figure 4.6 is a graphical illustration of the condition of a one-point stimulation applied to a fingertip. Specifically, this figure serves as a model for the spread of tactile sense perception and of its suppression that latently exists. The positive side of the figure indicates the receptive field, whereas the negative side shows the inhibitory field. The vertical axis has an arbitrary scale. Although the research method differs, this model's shape and properties resemble the neural unit proposed by Békésy.

At this point, the neural unit will be tentatively referred to as the "spatial sensory unit." The concept of the spatial sensory unit will be useful for information compression on designing sensory substitutes as the difference of the maximum transmitted information would be estimated by comparing the shapes and properties among human senses. The author would like to propose "temporal sensory unit" that is analogy to the spatial sensory unit in order to compare the temporal characteristics of the different senses.

4.3.2 Representation of Temporal Characteristics Based on Psychophysics

Given that the intensity and frequency of speech and most surrounding sounds change over time, it is important to determine if sensory aids can keep up with the changes. From the viewpoint of temporal characteristics, two kinds of investigations have been conducted. One focused on the fusion time of two successive stimulations; that is to say, the temporal difference between two successive short sounds when subjects can be barely perceived as two distinct sounds. As the fusion time for two successive stimulations corresponds to a two-point threshold in space, the author provisionally calls this a "two-point fusion time". An example of the two-point fusion time is shown later in the case of some hearing impaired children. The other investigation involved masking characteristics, which constitute an increase in the threshold of signal sounds presented before or after a masker. This is called "temporal masking" that can be used to examine the temporal characteristics of the senses.

4.3.2.1 Comparison of Auditory Sense with Tactile Sense from a Viewpoint of Temporal Masking

By using amplitude modulation (AM) as a masker, it is possible to compare the time properties of different senses by using the temporal masking process; namely, a masking curve [6, 7].

In this experiment, as Fig. 4.7 shows, we repeatedly stimulated the subject through sound or vibration. In a separate process, we also overlapped short signal stimulations at different time intervals. It goes without saying that the signal threshold varies according to the overlapping point in time. By measuring the threshold of all masker processes, the masking curve can be found.

In the example of the auditory sense experiment, the masking level was measured by overlapping AM sounds with signal sounds of 3 kHz and 5 ms at an optional time. In the tactile sense experiment, two bone-conducted phone receivers

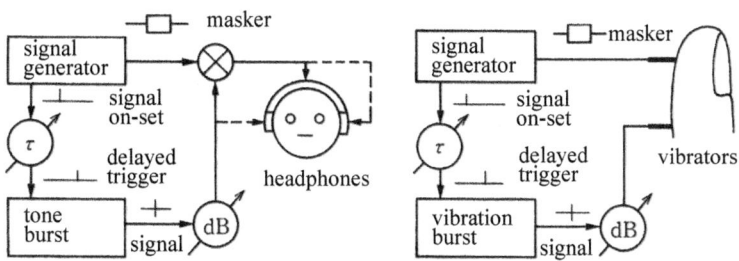

Fig. 4.7 Experimental method for temporal masking in auditory and tactile senses

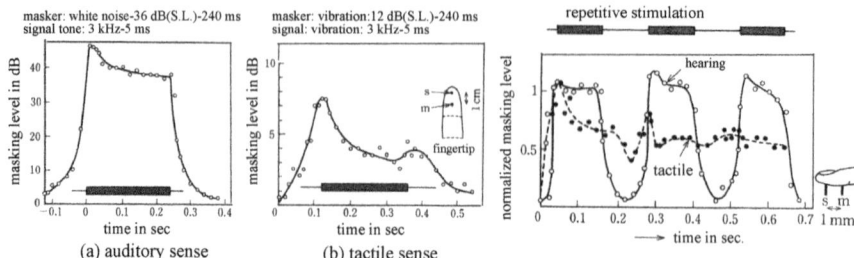

Fig. 4.8 *Left* Auditory temporal masking curve by a burst masker tone (240 ms). *Center* Tactile one by a burst masker vibration. *Right* Comparison of auditory temporal masking with tactile one by repetitive burst masker

were used to mask sounds produced by vibrators. By applying AM vibrations to one side and signal vibrations of 5 ms to the other side, the tip of the right index finger was stimulated by placing the vibrators 1 cm apart. The vibrator touched an area of 0.8 mm^2 and the diameter of the hole in the enclosing board was 3 mm. In the experiment using AM stimulation, rectangular waves with a frequency of 200 Hz were used as the carrier in the case of the tactile sense, and white noise was used in the case of the auditory sense.

As the left figure of Fig. 4.8 shows, the masking curve was determined using an AM stimulation, where the stimulation intensity changed with the shape of rectangular waves. Backward masking, which appears before the start of stimulation, was 25 ms in the case of the auditory sense, whereas in the case of the tactile sense, it was almost three times as long, which is almost the same result as shown in the center figure. At the time stimulation began, overshoot appeared in both cases. Moreover, at the beginning of the masking curve, both auditory and tactile senses showed similar properties.

However, as the right figure shows, when the burst-stimulation of the masker was repeated, the auditory masking revealed the same repeating pattern, whereas tactile masking varied considerably as time passed. This may occur due to the adaptation of both senses. It is clear that the tactile sense adapts quickly. Considering that pressure changes sharply for sounds emitted by the human voice, this phenomenon becomes a major problem when the tactile sense is used as a substitution for the auditory sense.

4.3.2.2 One Estimation Method of Temporal Characteristics in Hearing Impaired Children

Students with sensory neural hearing loss at a school for the deaf mainly use hearing aids to receive vocal sounds. However, in many cases they cannot clearly perceive the sound merely by amplifying the sound volume. Although it has been reported that this is due to low frequency resolution, temporal resolution of the auditory channel may also decrease the ability of sound perception [8–10]. Many researchers have reported on temporal masking, with some results indicating that there is no difference between defective and normal hearing, whereas others claim that there is. By approximating a sensory image and temporal masking curve precipitated by short sounds according to the mean of a normal distribution density function, the author evaluated the temporal properties of the auditory sense by using those parameters.

In the experiments, seven deaf school students aged from 11 to 15 and five subjects with normal hearing aged 10–13 participated in both studies. Left figure of Fig. 4.9 shows the results obtained from using the two-point fusion time. Here, we hypothesized that u represents the standard deviation, given that the sensory images of the short sounds lasting 3 ms fall within a normal distribution density function $f(t)$ as shown in the figure. Where, Δt indicates the interval between two sounds. In the figure, the horizontal axis is Δt and the vertical axis is the percentage at which the subjects reported that the two sounds were perceived separately. The solid line in the figure shows the probability curve $P(t)$; the value of which is 1 minus the sum of two normal distribution density functions $f(t)$ when t is 1/2. From the figure, it can be understood that the above assumption regarding u almost holds true as the measured value and probability curve closely coincide with each other. Thus value u, which indicates a two-point fusion time when the probability of answering correctly is 50%, is estimated from the probability curve $P(t)$ by fitting the curve to the data shown in the upper graph. This estimation process is the same as mentioned in the previous section regarding a two-point threshold.

Differing from past data acquired through such means as gap detection, the average value of u, which indicates the temporal resolution of defective hearing, is 18.3 ms [11, 12]. This is not very different from the 17.9 ms for normal hearing.

The right side figure of Fig. 4.9 shows the results for forward masking. Based on nine hearing-impaired subjects, the solid line in the figure is an approximation of the average of the masking curve calculated by the exponential function $A \cdot \exp$

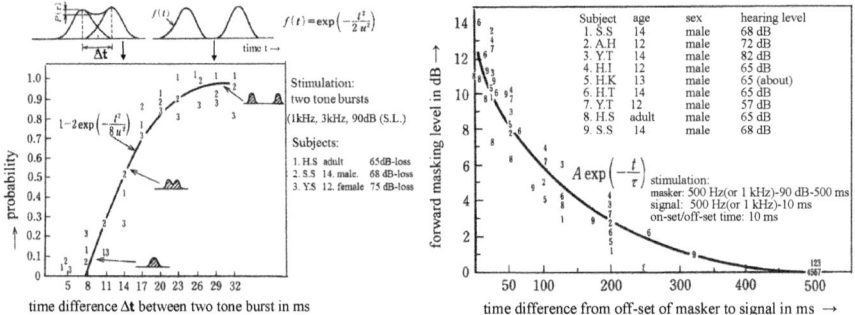

Fig. 4.9 *Left upper* Probability that two tone bursts were separately perceived. *Left lower* Sensory images produced by two short sounds. *Right* Forward masking pattern for hearing-impaired subjects

$(-t/\tau)$ shown in the figure. In the same figure, t indicates the time from the moment the masker finishes until the signal starts, while A is the assumed value of the quantity of masking (dB) when t is 0. Value τ represents the range that the masking covers. From the result of the examination, the author and his colleagues first noticed that the average value of τ for the hearing impaired in forward masking is considerably large at 108.4 ms.

Although there is a difference between trained and untrained students who are hearing impaired, for those who possess normal hearing the value of τ is within 30 ms. After examining backward masking for subjects with defective hearing and normal hearing, the subjects approximated its value using an exponential function, as in the case of forward masking. The constant τ for defective hearing is about 8.7 ms, whereas for normal hearing, it is about 8.2 ms, which does not constitute a very substantial difference. It is expected that for these hearing-impaired subjects, merely amplifying vocal sounds is not sufficient to raise speech articulation. Thus, it will likely become necessary to process the sounds so as to decrease forward masking. Hearing aids, possessing the same characteristics of differential calculus, might be one solution for decreasing the masking that is caused when vowel sounds are combined with consonants, thus amplifying radical sounds like consonants.

The reason two sounds are perceived as being merged is because the sporadic impulse discharged through stimulation by short sounds is continuously perceived until the perception of the following sound impulse, thereby resulting in the sensation of only one sound. Conversely, temporal masking indicates a temporal change in sensitivity to the signal sound before the masker is introduced and after the masker is removed. Thus, it is difficult to represent these two different phenomena by using only one scale.

4.3.2.3   One Representation Method of Temporal Characteristics

However, by thinking the same way as a two-point threshold, it is possible to calculate the time difference between the two stimulations when they are perceived as stimulations that are released successively. Although at this point it is not necessary to go into detail, it is worth noting that the time difference in the case of the auditory sense is about 5 ms, while for the tactile sense at a fingertip it is about 20 ms.

Figure 4.10 shows the pattern wherein the concept of a neural unit of the tactile sense is adopted by corresponding the two-point fusion time with the two-point threshold, and spatial masking to temporal masking. The plus side represents the zone that is temporarily added, while the minus side represents the zone that is suppressed. As shown inside the figure, the values u and τ of the auditory sense are indicated for comparison, so the difference between these unit scales can be intuitively understandable. By comparing the values u and τ of the auditory sense with those of the tactile sense, it is assumed that the tactile sense is around three times poorer at temporal resolution. In any case, the representation shown in the figure is

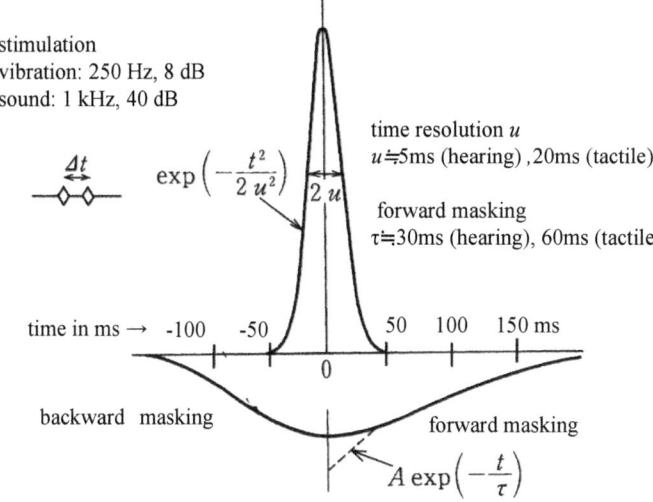

Fig. 4.10 Schematic representation of temporal characteristics of the tactile sense (comparison data of time resolution u and masking area τ of the auditory sense with the tactile sense are shown *inside the figure*)

helpful for intuitively grasping the temporal characteristics of senses as well as for providing useful information for the design of sensory aids and sensory support devices.

4.3.2.4 Temporal Characteristics of the Auditory and Tactile Senses Estimated from FM Masking

The transition part of the formants of consonants is a kind of frequency modulated (FM) tone. In order to investigate how the FM tone is emphasized in the auditory nervous system, the author measured the masking level of short sounds (signal) applied to subject's right ear, whilst exposing a subject's left ear to FM sounds (masker). The frequency of the FM masker sound was transformed by means of a ramp function and the modulation time was fixed at 20 ms. The masking level of the time was measured when the frequency was changing. From the experimental results, it was found that a sudden overshoot of masking level appeared as shown in the left figure of Fig. 4.11.

When the relationship is determined between this peak value and the rate of frequency change, it can be seen that the peak value of the masking level increases as the frequency increases, as the solid line shows in the right of Fig. 4.11. The transition parts of the formant as consonants change to vowels at about 10–100 Hz/ms and are characterized by the changing sound frequencies. As is obvious from the figure, the masking level increases remarkably when the frequency changes in this range. Thus, from the fact that FM sounds significantly

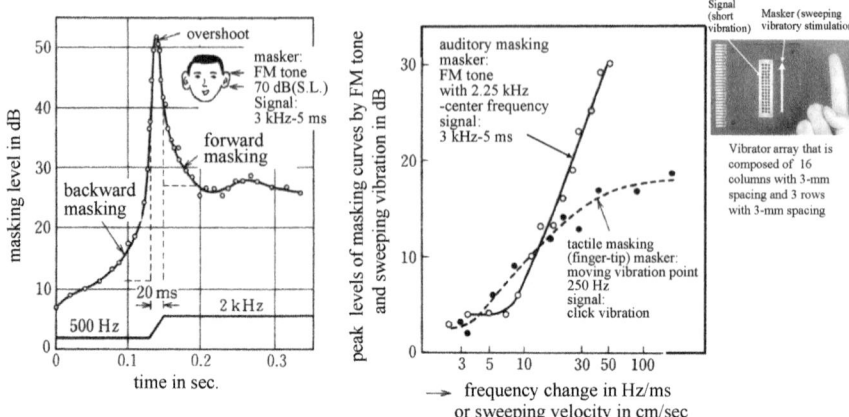

Fig. 4.11 *Left* An example of masking pattern by FM tone. *Right* Peak level of auditory masking by FM tone (*solid line*) as a function of frequency change and peal level of tactile masking (*broken line*) by sweeping vibratory stimulation as a function of sweeping velocity

mask other sounds, it is assumed that the auditory system contains a mechanism that emphasizes the transition part of the formants of consonants.

Conversely, as regards the tactile sense, after measuring masking by FM stimulation, whose frequency changes, no increase was found in masking as the frequency changed. This also suggests that the tactile sense does not have a mechanism of feature extraction for vibrating stimulations whose frequency changes. At the level of hair cells, which are made up of auditory receptors, FM sounds stimulate hair cells successively in a scanning manner. Accordingly, in the tactile sense experiment, we also measured tactile masking. By using a vibrator array arranged in 16 lines, each 4 mm apart, we created vibrating stimulations whose vibrating point moved in succession. Those stimulations were applied to the entire right index finger to measure tactile masking, using a vibrator array shown inside the right figure of Fig. 4.11.

The relationship between the movement speed of the vibrating point and the masking level is shown by the broken line in the right of Fig. 4.11. It was found that the masking level became extremely large when the movement speed of the vibrating part was 5–50 cm/s and the masking curve reacted the same. When the vibrator array was applied to the finger, the moving vibrating stimulations were clearly felt—as if the finger was being rubbed or stroked. This implies that both feature extraction for FM sound of the auditory sense as well as feature extraction for rubbing stimulation of the tactile sense follow a similar mechanism.

This enables us to say that it is preferable to convert FM sounds, which are characteristic of consonants as well as surrounding FM sounds, into moving vibrating stimulations. It is also evident that the method of densely arranging the vibrators to convert the pitch frequency of sounds into vibrating points is effective.

4.3.3 Estimation of Mechanical Impedance of Skin for Designing Sensory Substitutes

It is necessary to develop a tactile array or a 2D (two dimensional) tactile display that may transmit a speech spectral like a basilar membrane inside the cochlea. In previous research, Pawluk and his colleagues developed a tactile display composed of small solenoids with flat frequency characteristics up to 100 Hz [13]. However, considering the fact that the tactually perceivable bandwidth of the human skin is up to 500 Hz at most, it is preferable to provide tactile stimulus in a wider frequency range. Additionally, Summers and Chanter developed a tactile display using piezoelectric bimorphs that can provide tactile sensations up to 400 Hz [14]. However, the actuators in their device had a resonant frequency of 140 Hz when vibrating fingertip skin. Consequently, the display did not seem capable of completely controlling the parameters—phases, frequency components, and waveforms—of the tactile stimuli. Moreover, so as to provide more precise tactile stimulation, the design of the actuator should be determined in consideration of not only the actuator's behavior but also the mechanical dynamic characteristics of the skin of a finger.

To solve these problems, Homma and his colleagues measured the mechanical impedance of fingertip skin during use of a tactile display. From the measured impedances shown in the left figure of Fig. 4.12, a mechanical simulation model of the piezoelectric bimorph was also proposed when it is attached to the skin of a fingertip. The performance of the actuator was examined in experiments carried out with a piezoelectric actuator vibrating the skin of a fingertip. As full details of the simulation model are provided in their paper [15], here only a brief overview will be given. First, based on their measurements, the mechanical impedance of a fingertip surface ($Z(\omega)$) at a vibratory angular frequency of (ω) is expressed in the form:

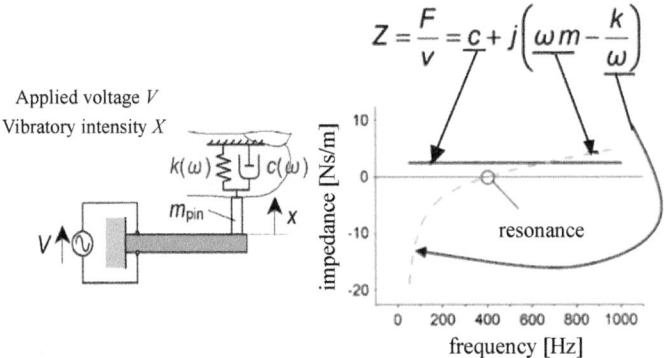

$$Z = \frac{F}{v} = c + j\left(\omega m - \frac{k}{\omega}\right)$$

Fig. 4.12 *Left* Equivalent circuit model of bimorph attached to surface of fingertip. *Right* Mechanical impedance $Z(\omega)$ of surface of fingertip and its frequency characteristics

Next, using an equivalent circuit model of the piezoelectric bimorph developed by Cho and his colleagues [16], the equivalent circuit model of the bimorph can be seen in the left part of Fig. 4.12. In this circuit, $Z(\omega)$ denotes the mechanical impedance of the fingertip skin, and m denotes the mass of the contactor pin. Finally, the vibratory displacement (x) of the fingertip skin can be obtained as an integral of the electric current shown in the right upper equation of the figure.

From the measurement, real and imaginary parts of mechanical impedance as function of vibratory frequency as shown in Fig. 4.13. The result indicates that no resonant occurs in the range between 0 and 1000 Hz and parameter C nonlinearly changes in accordance with vibratory frequency (ω). Therefore, the mechanical impedance can be represented by Eq. 4.1. As a frequency range of up to about 400 Hz is important for the transmission of information through vibro-tactile senses, it is desirable that both the actuator's frequency-displacement and frequency-phase characteristics are kept flat within the preferred frequency ranges as shown in the figure. Furthermore, as the tactile absolute threshold increases in a lower frequency range up to about 50 μm (p-p), it is also preferable that the actuator can provide greater vibratory displacement than 50 μm (p-p) throughout the entire tactilely perceivable frequency range.

$$Z = c(\omega) + k(\omega)/j\omega \qquad (4.1)$$

where $c(\omega)$ and $k(\omega)$, respectively denote the viscous coefficient and the elastic coefficient, which vary depending on angular frequency, while j denotes the imaginary unit.

In view of the above, Homma and the author established two guidelines in our design of the piezoelectric actuator. The first is to set the resonant frequency of the actuator to more than 1000 Hz as well as to keep the phase lag at 1000 Hz below 30° when vibrating a fingertip. The second is to provide vibratory displacements to the fingertip at more than 50 μm (p-p) in the range of 0–1000 Hz. We also designed new piezoelectric actuator that satisfied the above guidelines as described at the end of this chapter.

Fig. 4.13 Real and imaginary parts of mechanical impedance as function of vibratory frequency

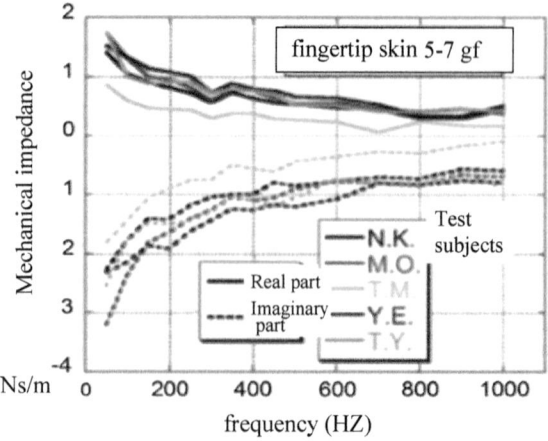

4.4 Tactile Aids for the Deaf and/or Blind

4.4.1 Conventional Tactile Aids

The idea of conveying sound information by means of vibrating stimulation through the skin was given birth in 1955 by Békésy [see 5]. Based on his ideas, several kinds of tactile aids were developed in the 1960s–1970s. In 1966, Suzuki and his colleagues designed a device that divided the frequency range of vocal sounds into 10 segments. The device vibrated 10 voice coils according to the intensity level of each frequency component, and applied the vibration to the 10 fingers of a subject [17]. It was named the TACTPHONE, and the results of experiments on it have yielded a lot of information, such as the fact that the tactile sense can serve as an aid to lip reading. However, the research on this device was not put to practical used due to the impracticality of its necessitating the use of all 10 fingers.

Spens designed a tactile aid device to transmit pitch information to the fingertips by using a vibrator matrix (6 × 24) which is used in a reading device called the OPTACON for the blind [18] as shown in the left of Fig. 4.14. The device extracts a sound's pitch, which causes certain rows of the vibrator matrix to vibrate according to the pitch's frequency. By adopting a method that sweeps from left to right over the vibrating point—much like an electric bulletin board—the most suitable sweeping speed was determined, based on voice-recognition experiments.

Sparks and Saunders adopted a method of placing an electrode matrix onto the abdomen and presented spectral patterns through electrical stimulation [19]. As far as sound pressure intensity is concerned, the former converts it into a number of stimulating points, whereas the latter converts it into stimulation intensity. The

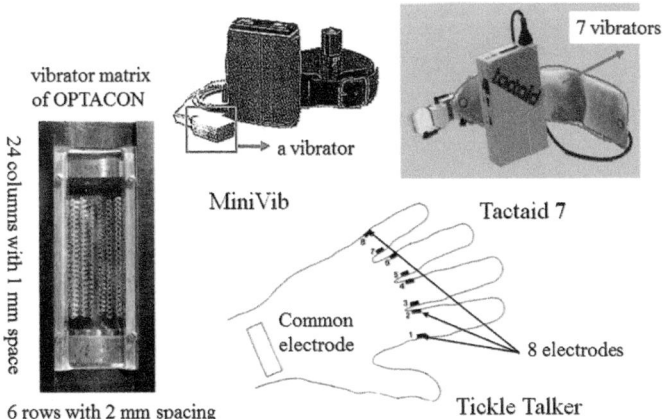

Fig. 4.14 *Left* Vibrator matrix (6x24) used for OPTACON. *Right* Three examples of tactile displays: MiniVib, Tactaid 7 and Tickle Talker

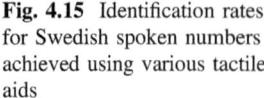

Fig. 4.15 Identification rates for Swedish spoken numbers achieved using various tactile aids

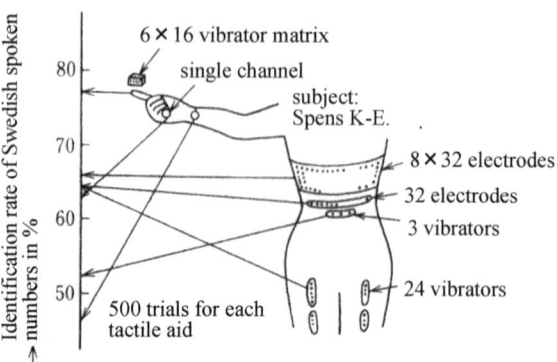

Tickle Talker [20], which was developed in Australia by Clark and his colleagues, uses a method of sticking two electrodes onto the base of four fingers. It has been shown that the recognition of consonants improved in the range of about 40–80% when this method is used in conjunction with lip reading. Another unique method, called the Tadoma method, has also been eagerly researched [21]. This method consists of fixing sensors to the area around a speaker's mouth to detect the movement of the mouth and thereby convey speech to the hand of a deaf person by way of tactile stimulation. The right side figures of Fig. 4.14 indicate three examples of the tactile display that were put to practical use in the 1970s.

By wearing various tactile aids that were being researched in many countries around 1976, Spens himself became a subject in his experiments involving the recognition of the speech of people who were pronouncing Swedish numbers. The result indicated that the fingertip is the most appropriate part for conveying speech signals and that vibratory stimulation using a vibrator array is superior to other stimulation methods such as electrical stimulation and single vibratory stimulation, as shown in Fig. 4.15.

4.4.2 Design and Efficacy of the Tactile Vocoder

Around 1974, the author conducted basic research to determine which sense is most suitable as a substitute for the auditory sense and to ascertain how the auditory sense differs from the tactile sense. Following this research, the author completed the development of a device called the tactile vocoder, by displaying vibrating patterns to one finger [22]. As Fig. 4.16 shows, the basic procedure is to first receive vocal sounds through a microphone, and then to pass them through an electric filter that serves as an external ear. External ears are a kind of wide-area filter with the unique characteristic of increasing or widening the range that receives vocal sounds. Through this circuit, the sound is relayed to a filter bank whose function is to divide the frequency range from higher to lower parts, just like

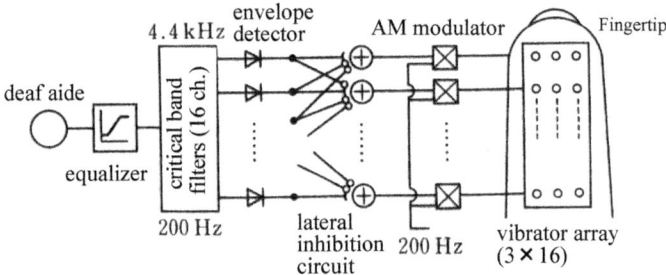

Fig. 4.16 Block diagram of tactile vocoder with a vibrator array (3 × 16)

internal ears. This dividing method is based on a filter bank called a critical band width composed of 24 band-pass filters, which is ascertained to exist in the auditory system. Actually, 16 band-pass filters, which simulate a critical band width (200–4400 Hz), are installed in the tactile vocoder. These 16 channels correspond exactly to the range that contains the components of the first and the second formants, and some consonants.

As Fig. 4.17 shows, envelopes are detected from the wave shapes of each output of the filter, after which they are sent to the lateral inhibition circuit. The function of this lateral inhibition circuit is similar to the lateral inhibition nervous circuit of the auditory sense. Specifically, it plays a role of emphasizing formants and stressing weak sound components like consonants. It is known that the neural inhibition of

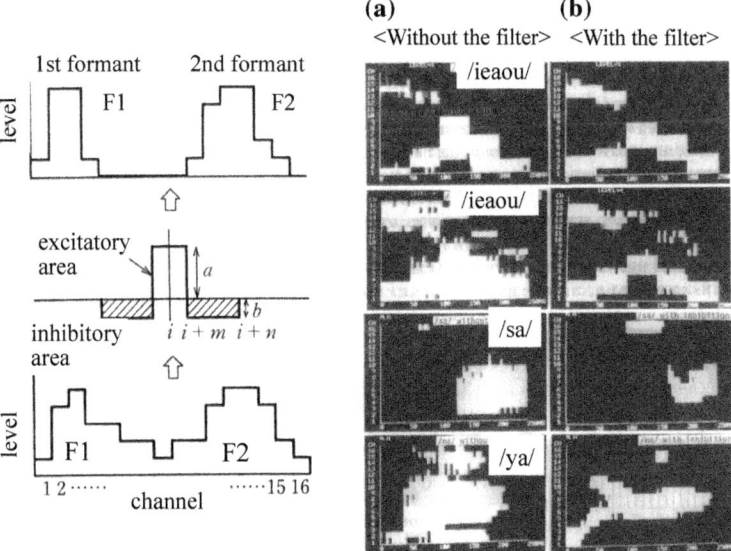

Fig. 4.17 *Left* Function of spatial band-pass filter modeled after lateral inhibition of tactile sense. *Right* Speech spectral patterns processed by spatial band-pass filter (*a* without the filter, *b* with the filter)

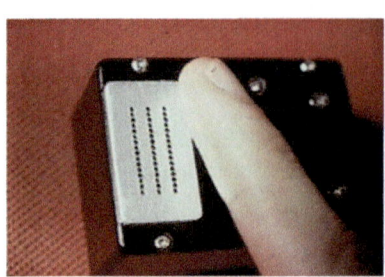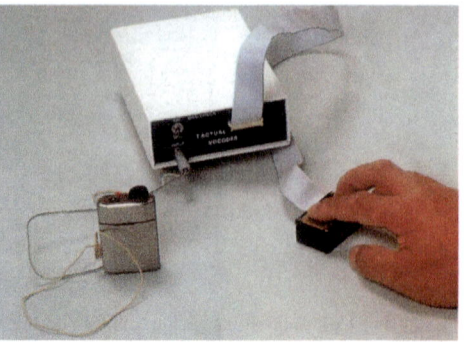

Fig. 4.18 *Left* Vibrator array (3 × 16). *Right* Wearable tactile vocoder manufactured in Japan

the tactile sense is more diffused compared to that of the auditory sense. The author used a sharpening function to compensate for the low tactile spatial resolution.

The author calculated the parameters of the lateral inhibition circuit, where five vowel sounds of the Japanese language can be distinguished most clearly and distinctly, as shown in the left figure of Fig. 4.17. He found that the temporal spectrum patterns of the five vowel sounds show almost the same patterns, and are not dependent on the normal sound level. Furthermore, the formant transition points of the consonant sounds /sa/ and /ya/ are clear after they pass through the lateral inhibition circuit shown in the right figure.

Piezoelectric elements were driven so that the vibrations could be received by the tactile sense, especially by the Pacinian corpuscles, and the outputs were converted into vibrations of 200 Hz. As the left side of Fig. 4.18 shows, these elements are arranged into 16 rows 1 mm apart and three columns 3 mm apart, resulting in a total of 48 elements. By fixing brush pins on the top of the piezo-electric elements, the output was converted into vertical vibrations. They were lined up 1 mm apart, based on the result that 1 mm is the minimal distance necessary to detect the movement of a vibration point. Actually, the tactile vocoder was designed to be able to detect the formant transition of consonant sounds. The reason the piezoelectric elements are arranged in three columns is to make the vibrations clearly recognizable. Three millimeters almost corresponds to the two-point threshold necessary to detect two stimulations. The right figure shows a photograph of the manufactured tactile vocoder.

As can be seen in the left figure of Fig. 4.19, when the author measured the identification rate of vowels by using other part of the body, such as the palm, the differences among the body parts become clear. For all of the subjects (five adults with normal hearing), the identification rate reached the highest level when using a fingertip [23]. This result can also be expected when conducting basic experiments; however, the difference in vowel-recognition ability seems to particularly reflect the difference between two-point thresholds. This result also confirms that the fingertip is the most suitable part of the body for use as a substitute sense.

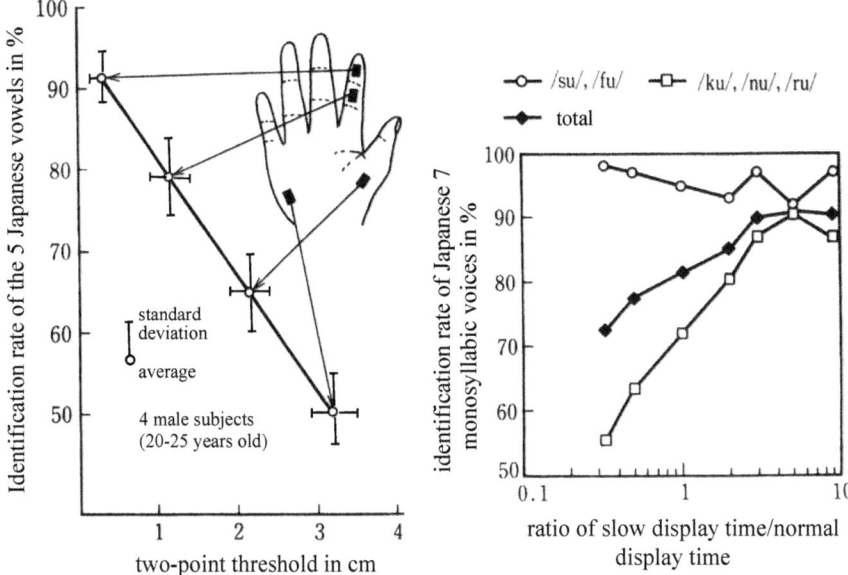

Fig. 4.19 *Left* Identification rate for five Japanese vowels as function of stimulating part. *Right* Increase in consonant identification by slowly displaying consonant part onto fingertip

This method clearly facilitates lip reading, a process that consists of reading the lip movements of another person to understand what they are saying. For example, the author and his co-researcher asked hearing-impaired subjects to use lip-reading in order to tell us which consonants people had uttered, after the speakers had randomly pronounced the following seven consonants: /ku/, /tsu/, /su/, /nu/, /yu/, /ru/, and /fu/.

The subjects could correctly distinguish only about 27% of the sounds. However, after undergoing training in the use of a tactile vocoder combined with lip reading for one hour a day during a period of one week, the rate of correct consonant identification increased to 70%. As this device was proven to be an effective aid to lip reading, it was later manufactured as a product for use in various research facilities.

However, it became clear that plosive speech sounds such as /p/, /t/, /k/ are difficult consonants to distinguish, even after extensive training. Basic research aimed at improving the ability to discriminate such sounds was continued. One reason it is difficult to discriminate such plosive sounds is because their duration is very short and they start abruptly. In point of fact, when presented more slowly by means of a computer, the correct rate of discrimination for the sounds /ku/, /nu/, /ru/ rises sharply because the sounds are lengthened by a factor of 3–4 as shown in the right figure of Fig. 4.19. As mentioned in the previous session, this result is also supported by the fact that the tactile sense is about three times lower in time resolution than the hearing.

Wada and his colleagues reported that consonant sounds, particularly plosives such as /pa/, /ta/ and /ka/, can be distinguished very well when the patterns are presented in a horizontal sweeping motion as on an electric bulletin board. They found that the correct identification rate peaks when the sweeping speed is about 10 cm/s by using a vibrator matrix of 16 columns and 4 rows with a vibrator spacing of 1 mm [24]. This can be explained by noting that with a conventional tactile vocoder, in the case of consonants like plosives (/ka/, /ta/, /pa/) as a vowel comes immediately after a consonant on a fingertip, the first consonant coming is backward masked by a vowel. Thus it is obvious that the backward masking is decreased by converting the temporal patterns into spatial patterns.

By applying the above findings, the author initiated experimental training for voice-sound recognition. Four hearing-impaired subjects (all 4th graders) were first made to learn the vowels /a, i, u, e, o/. Then, they were taught disyllabic vocal sounds made from /a, i, u, e, o/, such as /uo/, /ie/, /ao/. The subjects were then asked to memorize some names and vocal sounds that are often used on the telephone, including such words as "yes," "no" and "again." Next, the author conducted communication experiments with the subjects' parents by using actual telephones. The purpose of this experiment was to determine how well the students could communicate with the words they had learned while making phone calls to their parents. The record of these findings was broadcasted in a documentary TV program produced by NHK (Japan Broadcasting Association) in 1975, which is still considered a valuable document today.

4.4.3   Association with Language Cortex

People have the ability to separate each successively spoken word sound by sound by using a method called "segmentation." However, it is still impossible to conclude if the same process can be accomplished by the use of the tactile sense. Accordingly, it is necessary to develop a segmentation technique with a computer that can break up the various sounds and convert them into separate stimulations, after which, these vibrations can then be relayed to a fingertip.

Another problem is that the spectra of words differ greatly depending on each speaker. Human ears are able to recognize the same words even when spoken by different speakers. This is because the auditory sense possesses an ability to categorize speech sounds. Therefore, it is also necessary to develop technology that can perform segmentation automatically as well as normalize spectral changes resulting from differences in pronunciation between speakers. In order to create tactile aids that can aid in the formation of language concepts, similar to Braille for the blind, a major breakthrough is necessary.

However, encouraged by the neuroscience results regarding the plasticity of the neural network in the human brain as mentioned at the beginning of this chapter, the author designed a new model of the tactile vocoder for the deaf-blind in 2006, as will be discussed later in this chapter. Although many challenges remain, even the tactile aids that exist at present can help to facilitate the detection of human voices or of such sounds as a car approaching from behind. As described in the next section, tactile aids can be used for vocalization training so that, as an example, the hearing impaired can get feedback by way of the tactile sense when attempting to utter their own words in such a way as to match the words spoken by another person.

4.5   Vocal Training via Tactile Sense

Vocalization is characterized by such elements as the tongue, jaw and lips, which are structural characteristics that can be recognized through the visual sense. It has been some time since traditional training for an "oral method" has been developed by means of visual displays. However, as the intensity and pitch of vocal sounds are mainly controlled by the muscles of the vocal cords, it is difficult to visually detect structural changes in the vocal organs. This is especially true in the case of deaf children who tend to talk at an extremely high or low pitch, resulting in a distinct decrease in clarity of vocalization. As such, assistive aids for vocal training can be best applied to voice pitch training. At this point, the author would like to introduce a certain research methodology, based on control engineering, by describing some experiments involving pitch control through the tactile sense [25].

Figure 4.20 shows an experimental system for pitch control using the tactile sense consisting of a microphone, a computer and a vibrator array. In this system, a microphone is placed near a person's mouth. Vocal sounds received by the microphone are transmitted to the preprocessing section, which extracts the sound pitch and intensity. Next, the sounds are sent to an A-D (analog-to-digital) converter, then to a computer, the D-A (digital-to-analog) converter and finally, to the vibrator array. On the vibrator array there are 48 high-density piezoelectric elements. This is the same as the setup used for the vibrator array of the previously developed tactile vocoder.

In the display method, the pitch height matches the 16 rows of the vibrator array, while the sound pressure corresponds to the vibratory intensity. The upper part of

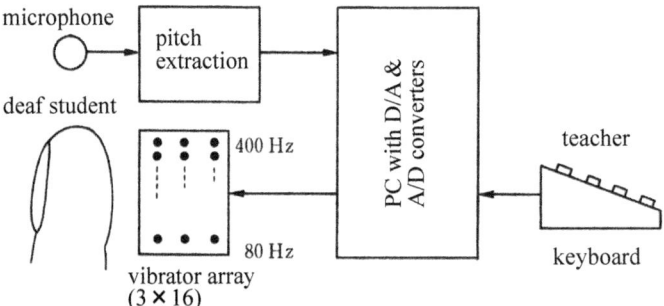

Fig. 4.20 Voice pitch control method by tactile feedback

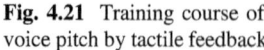
Fig. 4.21 Training course of voice pitch by tactile feedback

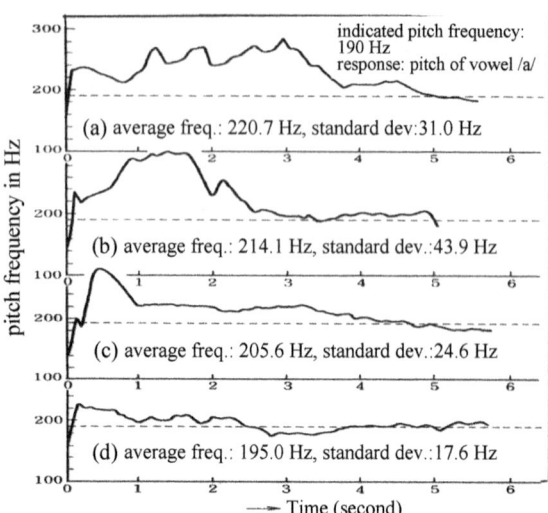

the finger is stimulated by higher pitches and the lower part by lower pitches. In the case of designating a target pitch, we alternated the presentation of the target pitch with the voice pitch on the vibrator array, and determined the difference between the two as the difference between two rows of a vibrating part. The results showed that pitch control through the tactile sense varied according to the characteristics and ability of the deaf children, although training the children to stabilize the pitch proved to be relatively easy.

Figure 4.21 shows an typical example of the results of pitch control conducted by a deaf student (12 years old). The pitch frequency of 100–400 Hz matched the vibrator array and the vowel sound /a/ was vocalized so that it would coincide with the target pitch of 190 Hz (indicated by the chained line). When the boy vocalized freely, the average pitch was found to be approximately 200 Hz. This experiment was repeated four times, at intervals of 10 min each. Figure 4.21a–d represent the results in the order of training. Specifically, in the case of (a), it took 5 s to reach the target pitch, 3.2 s for (b), and only 1 s for (d) and the number of large overshoots gradually decreased. Thus, the learning effect could clearly be observed even during such a short period.

During the course of this experiment, the author also examined the way in which the subject adjusted his voice pitch when the target pitch was increased by small increments. The experimental data are equivalent to the step response in control engineering. The purpose of this experiment was to analyze the tactile pitch control systems from the point of view of control theory. The target value was fixed lower than the natural average pitch frequency. Two to three seconds after presenting the stimulations, the target pitch was then raised step by step in increments of 50 Hz. The subject who took part in the experiment had completed training under conditions of a fixed pitch frequency.

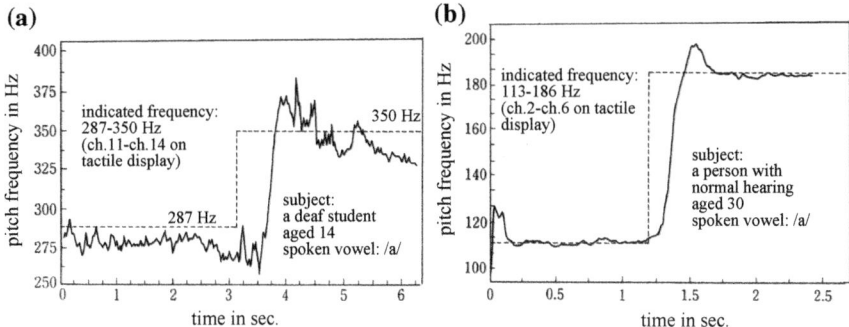

Fig. 4.22 An example of response of pitch frequency controlled by a deaf student. **b** That by a student with normal hearing

Figure 4.22a shows representative examples of three deaf students whose hearing ability loss was greater than 90 dB. Figure 4.22b shows an example of a person with normal hearing, while the dotted line indicates the target pitch.

From this experiment, it was found that the voice pitch usually started up about 0.2–0.3 s after the target value changed and an overshoot appeared several 100 ms later. However, by repeating periodic changes, the voice pitch approached the target value. As a comparison with normal hearing, the figure Fig. 4.22b was added, with this case also showing a time delay and overshoot as in the case of deaf student. Naturally, in the case of normal hearing, once the pitch coincided with the target value, auditory feedback could be used. Consequently, accurate pitch control was achieved in a short time. This time delay for subjects with normal hearing almost coincided with the above-mentioned vocal reaction time. Even this simple problem of pitch control reflects a basic idea regarding the evaluation of sensory aids for vocal training.

The block diagram in Fig. 4.23 shows a vocal control system using sensory aids. As Fig. 4.23a shows, since subjects with normal hearing can receive both their own pitch pattern as well as the target pitch pattern through the auditory sense system, they can compare them. However, as Fig. 4.23b shows, for a deaf student to accomplish the same thing, they require a sensory aid system P(s) combined with a tactile system T(s). From the viewpoint of control engineering, the purpose of this research is to determine the P(s) that will enable the target pitch pattern X(s) to coincide with one's own pitch pattern Y(s) in the shortest period of time. Thus, it is important to conceptualize the whole system as one man-machine interface.

This section reports research regarding the pitch control of vocal sounds that attempts to introduce the above-mentioned idea to the maximum extent possible, and it can be regarded as helpful for research in developing training aids in addition to devices for pitch control.

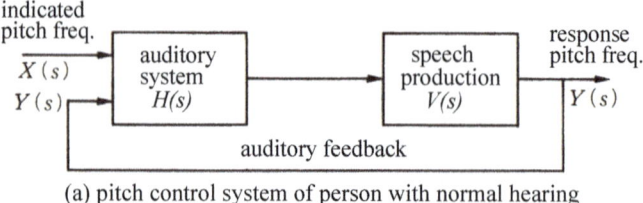

(a) pitch control system of person with normal hearing

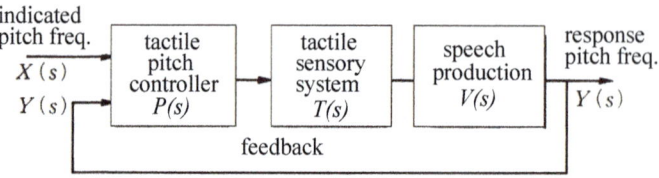

(b) pitch control system with tactile feedback

Fig. 4.23 Block diagram of pitch control system with/without tactile feedback

In 2002, the author moved to the University of Tokyo to work on a barrier-free project. The project was conducted by Professor Satoru Fukushima (Fig. 4.24a), who became blind at the age of 9 and deaf at 18. From the viewpoint of the deaf-blind, he suggested how we should perform barrier-free research. He usually communicates with us using both the tactile sense of six fingers through the help of interpreters. This tactile communication method is called "finger Braille," which Professor Fukushima invented together with his mother.

As shown in Fig. 4.24b, it is anticipated that finger Braille patterns would reach the visual cortex through the tactile cortex and then the patterns would be associated

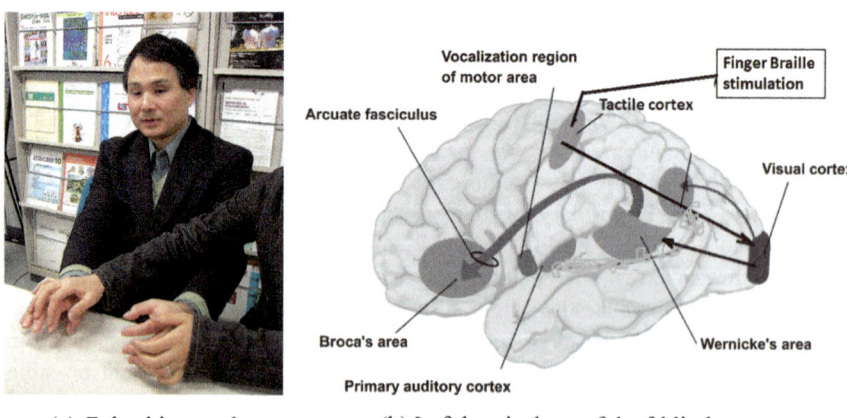

(a) Fukushima and   (b) Left hemisphere of deaf-blind person
"finger Braille"   while using finger Braille (hypothesis)

Fig. 4.24 a Fukushima and finger Braille that he invented. **b** Hypothesized information route in his left hemisphere while using the finger Braille

with the language understanding cortex. Based on the plasticity of the neural network in the human brain, lost functions might be compensated for by another sensory cortex.

Sakajiri and his colleagues designed a new pitch control device that consisted of a digital signal processor (DSP) and a piezoelectric vibrator array, as shown in Fig. 2.25a, so that the signal processing could be changed only by software [26]. For the deaf-blind, the new device was produced to transform a voice pitch frequency into a musical scale that corresponds to the vibrating point of a fingertip.

The subject was a woman who had been deaf-blind since she was 40 years old and was 67 years old at that time (2006). As she was a teacher of Japanese musical instruments and folk songs until she lost her visual and auditory senses, it was expected that she could easily handle the tactile pitch display and sing songs by using the feedback through her fingertips. After she learned the musical scale using the tactile device, she was able to sing some Japanese songs (Fig. 4.25b). This example shows that tactile information might help provide feedback for melodies [27]. However, congenital deaf-blind people were not able to use this technology to sing.

These findings lead to many suggestions as to whether tactile information could be transmitted to the auditory cortex and speech-understanding area in the brain, and helped us understand how brain and cognitive research are important in the design of assistive tools.

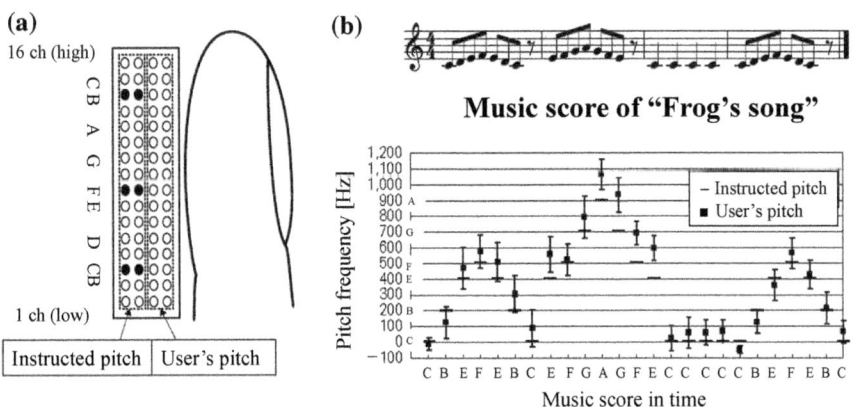

Fig. 4.25 **a** Vibrator display of newly designed tactile vocoder for pitch training. **b** Pitch pattern of deaf-blind person produced by a "Frog's song" where music score is presented on the *top of this figure*

4.6 Auditory Localization Aids

Another merit of substituting the tactile sense for the auditory sense is that it can provide information such as sounds that can warn of dangers outside the field of vision. With some improvement, this can even help to inform the user of the direction of a sound source. With that purpose in mind, by examining the mechanism by which the auditory sense determines the direction of a sound source, we can try to devise a way for the tactile sense to accomplish the same thing. It might also be necessary to examine to what degree the tactile sense can recognize the position of a sound source image as well as its movement.

Békésy and Alles showed that the tactile sense can perceive a difference in intensity between two tactile stimulations displayed on a palm. When a sound reaches both ears, the listener's ears can perceive the sound's direction based on the arrival time difference of the sound at each ear [28, 29]. Accordingly, to learn how precisely a sensory image can be recognized, Niioka and his colleagues provided a palm vibratory stimulation that has the same time difference and intensity difference as in the case of the auditory sense when sound reaches both ears [30].

The results indicated that using vibratory stimulation alone makes it more difficult to perceive the time difference between two stimulations compared to the case of the perception of time differences between sounds reaching both ears. That is, in the case of vibratory stimulation, the location of a sensory image (phantom sensation) is perceived only based on the intensity difference between two vibrating stimulation points. In short, by converting sound time differences into vibratory intensity differences, one would be able to perceive the location of a sensory image.

The just-noticeable level of a sensory image's movement can be determined by using amplitude modulation of two vibratory stimulation with two reverse-phase sine waves. The figures inside of the left and the right figures of Fig. 4.26 show the intensity of pulse stimulation changing like a sine wave. When the intensity difference is 0, the sensory image is located to the center, where the area of the phantom sensation is largest.

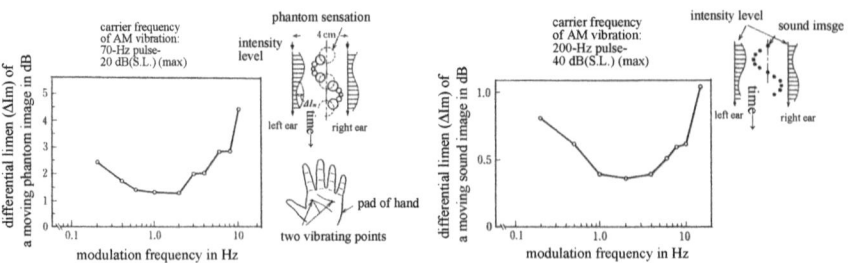

Fig. 4.26 *Left* Differential limen of moving phantom image produced by intensity difference of two vibration stimuli. *Right* Differential limen of moving sound image produced by intensity difference between right and left ears

Judging from the result in the left figure, the minimum shift-detection limit for the steady state caused by standing stimulation is 3.7 dB. This tells us that the shift-detection limit decreases when the sensory image moves. When it moves in the range of 1–2 Hz, the detection limitation is lowest and the movements of the sensory image can be detected with an intensity of 1.2 dB. This is the same as the result for the auditory sense as seen in the right figure. Moreover, both the auditory and tactile senses can follow the movement's shift frequency up to 20 Hz.

Based on the points mentioned above, Niioka and colleagues designed a device for locating sensory images by using the tactile sense. The basic purpose of this invention is to use the sounds reaching two microphones placed on both sides of the device to control the vibration amplitudes of two vibrators. The sounds that we usually consider important in our daily life, such as car sounds as well as vocal sounds, have a relatively low frequency that is in the 500–1000 Hz range. Taking this into account, in the case of determining the location of sensory images by the tactile sense, as the tactile sense cannot make use of any information regarding time difference it is necessary to convert that temporal information into an intensity difference that the tactile sense can perceive. More specifically, an envelope detector was used to extract the intensity of a microphone's output, which, in turn, modulates the amplitudes of low frequency impulses to activate the vibrators, as shown in the upper of Fig. 4.27.

We conducted an investigation to detect the direction of a moving vehicle. In order to collect sounds, two non-directional condenser microphones were used and placed horizontally at 1.3 m above the ground and 20 cm apart. As shown in the

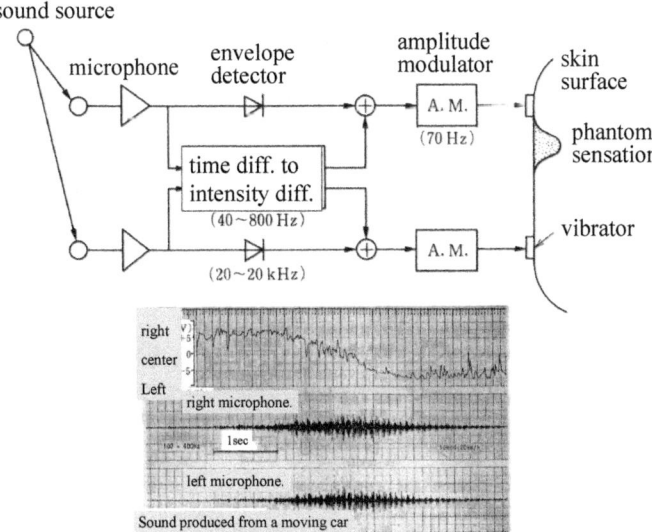

Fig. 4.27 *Upper* Block diagram of tactile sound localizer. *Lower* Intensity difference change of two vibratory stimuli obtained by tactile sound localizer, where sound source is a moving car

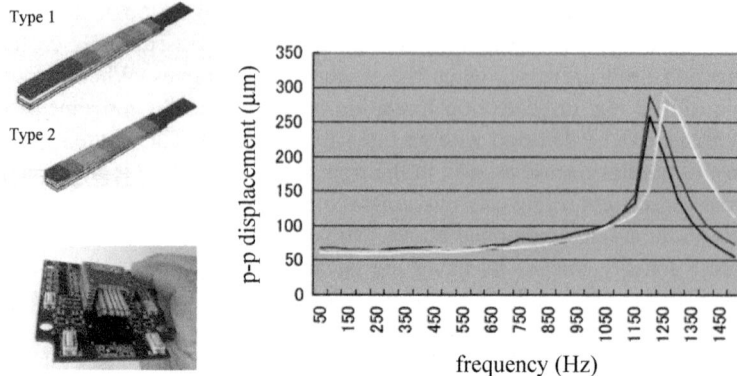

Fig. 4.28 *Left* Appearance of before multilayer-bimorph piezoelectric vibrator. *Lower left* 2D tactile display for mobile phones. *Right* Displacement of the vibrator versus frequency of applied voltage

bottom graph indicates the input waveform for the time-difference detecting circuit while the upper part of the graph shows the output waveform. The positive value of the voltage for the time difference detection shows that the arrival time for the right side was earlier than for the left side. This example was taken when a car passed at the speed of 20 km/h from right to left at a distance of about 2.4 m from the microphones as shown in the lower of Fig. 4.27.

The amplitude of a 70 pps pulse was modulated to create an intensity difference between the pulse vibrations of the two vibrators. When the two vibrators were placed 4 cm apart on a palm or 10–20 cm apart on a person's abdomen, chest, or

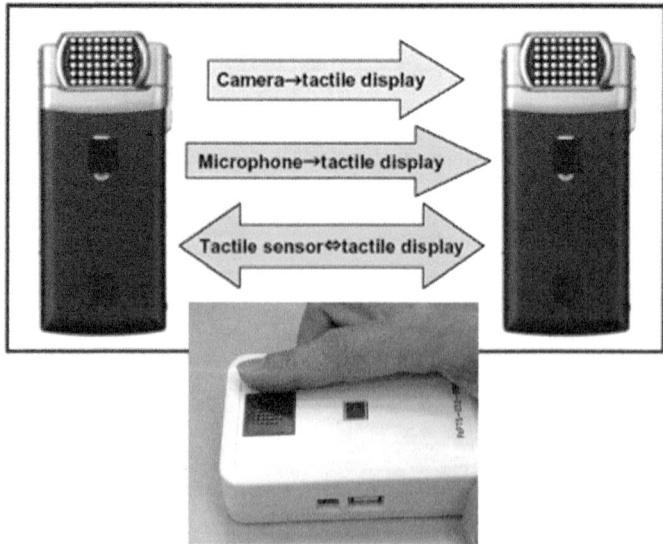

Fig. 4.29 Haptic display attached to mobile phone for auditory and/or visual impairment. It is also used for virtual reality display

back, it was confirmed that any area of the body could utilize the tactile sense to adequately perceive the movement and direction of an approaching vehicle.

Recently, the author and his colleagues improved Homma's actuator so it could be activated at a low voltage (<10 V) by using a multilayer-bimorph piezoelectric device that Kyosera LTD. developed, as shown in Fig. 4.28. The actuator has been used in a 2D tactile display for mobile phones and it was also applied to detect the surrounding sounds by moving the mobile phone. Furthermore, we designed a haptic display that can be attached to mobile phone for the auditory and/or visual impaired and also for a virtual reality display as shown in Fig. 4.29 [31].

4.7 The Future of Tactile Display

Tactile aids today mainly use vibratory stimulation. The tactile sense of human beings is not only for receiving vibratory stimulation but also for receiving a sensation to perceive the quality of a material. As the feeling of a material is something that can only be expressed qualitatively through words, the quantity of such information is incomparably larger than what can be conveyed through vibratory stimulation alone. Considering the fact that a human voice is recognized by the quality of its sound the invention of tactile aids that could convert such *sound quality* into *material quality* would dramatically increase the amount of obtainable information by easily combining tactile sense information with the quality of vocal sounds. As to a tactile display for presenting quality of materials, a research conducted by Ino and his colleague may serve as a useful reference [32].

For this reason, present-day research in the field of virtual reality is placing considerable effort on the development of a tactile sense display that can convey the feeling of a material's quality. Virtual reality-based research into the tactile sense might eventually lead to an entirely new concept in the development of auditory aids as well as visual aids.

References

1. N. Sadato, A. Pascual-Leone, J. Grafman, V. Ibanez, M.P. Deiber, G. Dold, M. Hallett, Activation of the primary visual cortex by Braille reading in blind subjects. Nature **380**(6574), 526–528 (1996)
2. S. Levanen, V. Jousmaki, R. Hari, Vibration-induced auditory-cortex activation a congenitally deaf adult. Curr. Biol. **8**(15), 869–872 (1998)
3. K.O. Johnson, G.D. Lamb, Neural mechanism of spatial tactile discrimination: neural patterns evoked by Braille-like dot patterns in the monkey. J. Physiol. **310**, 117–144 (1981)
4. S.J. Bolanowski, G.A. Gescheider, R.T. Verrillo, C.M. Checkosky, Four channels mediate the mechanical aspects of touch. J. Acoust. Soc. Am. **84**(5), 1680 (1988). doi:10.1121/1.397184
5. G. Békésy, *Sensory inhibition (Chapter II)* (Princeton University Press, Princeton, 1967)

6. T. Ifukube, Masking by frequency modulated tone. J. Acoust. Soc. Jpn. **29**(11), 679–687 (1973). (in Japanese)
7. T. Ifukube, Auditory masking of amplitude modulated tone and its analysis by analog simulation. J. Acoust. Soc. **31**(4), 237–245 (1975). (in Japanese)
8. D.A. Nelson, R.L. Freyman, Temporal resolution in sensorineural hearing-impaired listeners. J. Acoust. Soc. Am. **81**, 709 (1987). doi:10.1121/1.395131
9. E. Zwicker, K. Schorn, A.A. Ashoor, T. Prochazka, Temporal resolution in hard-of-hearing patients. J. Audiol. **21**(6), 474–492 (2009)
10. J.M. Festen, R. Plomp, Relations between auditory functions in impaired hearing. J. Acoust. Soc. Am. **73**, 652 (1883). doi:10.1121/1.388957
11. B.C.J. Moore, Frequency selectivity and temporal resolution in normal and hearing-impaired listeners. Br. J. Audiol. **19**(3), 189–201 (1985). doi:10.3109/03005368509078973Journal
12. P.J. Fitzgibbons, F.L. Wightman, Gap detection in normal and hearing-impaired listeners. Acoust. Soc. Am. **72**, 761 (1985). doi:10.1121/1.388256
13. D.T.V. Pawluk, C.P. van Buskirk, J.H. Killebrew, S.S. Hsiao, K.O. Johnson, Control and pattern specification for a high density tactile array, In *IMECE Proceedings of ASME Dynamic System and Control Division*, vol. 64 (1998), pp. 97–102
14. I.R. Summers, C.M. Chanter, A broadband tactile array on the fingertip. J. Acoust. Soc. Am. **112**(5), 2118–2126 (2002). doi:10.1121/1.1510140
15. T. Homma, S. Ino, H. Kuroki, T. Izumi, T. Ifukube, Development of a piezoelectric actuator for presentation of various tactile stimulation patterns to fingerpad skin, In *Proceedings of IEEE EMBS 2004*, vol. 2 (2004), pp. 4960–4963. doi:10.1109/IEMBS.2004.1404370
16. Y.S. Cho, Y.E. Pak, C.S. Han, S.K. Ha, Five-port equivalent electric circuit of piezoelectric bimorph beam. Sens. Actuators A-Phys. **84**, 140–148 (2000)
17. H. Suzuki et al., TACTPHONE as an aid for the deaf. in *Proceedings of 6th ICA* (1968)
18. J.G. Linvill, J.C. Biiss, A direct translation reading aid for the blind. Proc. IEEE **54**, 40–51 (1966)
19. D.W. Sparks, P.K. Kuhl, A.E. Edmonds, G.P. Gray, Investigating the MESA (multipoint electrotactile speech aid): the transmission of segmental features of speech. J. Acoust. Soc. Am. **63**, 246–257 (1978)
20. P.J. Blamey, G.M. Clark, Psychophysical studies relevant to the design of a digital electrotactile speech processor. J. Acounst. Soc. Am. **82**, 116–125 (1987)
21. S.J. Norton, M.C. Shultz, C.M. Reed et al., Analytic study of the Tadoma method: background and preliminary results. J. Speech Hear. Res. **20**, 574–595 (1977)
22. T. Ifukube, Maximum information transmission on a tactual vocoder: in the case of time invariant stimulation. Jpn. j. Med. Electr. & Biol. Eng. **17**(3), 230–236 (in Japanese)
23. T. Ifukube, H. Minato, C. Yoshimoto, Basic studies on tactile vocoder by psychophysical experiments. J. Acoust. Soc. Jpn. **31**(3), 170–178 (1975). (in Japanese)
24. C. Wada, S. Ino, T. Ifukube, Proposal and evaluation of the sweeping display of the speech spectrum for a tactile vocoder used by the profoundly hearing impaired. Trans. Insut. Electr. Inf. Commun. Eng. A, **J78-A-3**, 305–313 (1995)
25. T. Ifukube, Sensory prosthesis on speech training. Jpn. Ergon. Soc. **16**(1), 5–17 (1980). (in Japanese)
26. M. Sakajiri, S. Miyoshi, K. Nakamura, S. Fukushima, T. Ifukube, Accuracy of voice pitch control in singing using tactile voice pitch feedback display, in *Proceedings of IEEE SMC 2013* (2013), pp. 4201–4206
27. M. Sakajiri, S. Miyoshi, K. Nakamura, S. Fukushima, T. Ifukube, Development of voice pitch control system using two dimensional tactile display for the deaf-blind or the hearing impaired persons. NTUT Educ. Disabil. **9**, 9–12 (2011)
28. D.S. Alles, Information transmission by phantom sensations. IEEE Trans. MMS **1**(11), 85–91 (1970)

29. G. Békésy, *Sensory inhibition (Chapter III)* (Princeton University Press, Princeton, 1967)
30. T. Niioka, T. Ifukube, C. Yoshimoto, Basic studies of a tactual sound localizer for the deaf. J. Acoust. Soc. Jpn. **33**, ND.5 (1977) (in Japanese)
31. K. Yabu, M. Sakajiri, T. Ifukube, Development of a wearable haptic tactile interface as an aid for the hearing and/or visually impaired. NTUT Educ. Disabil. **13**, 5–12 (2015)
32. S. Ino, T. Odagawa, M. Sato, M. Takahashi, T. Izumi and T. Ifukube, A tactile display for presenting quality of materials by changing the temperature of skin surface, 1993. Proceedings of 2nd IEEE International Workshop on Robot and Human Communication (1993). doi.10.1109/ROMAN.1993.367718

Chapter 5
Speech Recognition Systems for the Hearing Impaired and the Elderly

Abstract In this chapter, after the author describes how various attempts to convert speech sounds into visual patterns have been applied to auditory substitutes, he introduces recent speech-recognition technologies that have been applied to captioning systems for the hearing impaired as well as the elderly. For supporting elderly people with cognitive decline, the number of which is rapidly increasing in a super-aged society, the author introduces an example of communication robots that give notification when it is time to take medication and remind elderly people of their daily schedule.

5.1 Visible Speech

There is a long history of auditory sensory aids that convey speech information via the visual sense. Such *visible speech* is a method of indicating sounds through pictures, and was first proposed by Potter et al. [1]. With this method, in the speech spectrograms shown to the hearing impaired, the horizontal axis of the pattern indicates time, the vertical axis indicates frequency, and the shading indicates intensity, as shown in Fig. 5.1.

At one time, research that utilized this method to *read* speech was very popular in the development of auditory sensory aids. The idea of substituting speech perception for the hearing impaired by means of the visual sense is based on the expectation that the visual sense's ability to receive information is more extensive than the other senses. It is certainly true that compared to the more limited tactile sense, such as that of a fingertip, the visual sense has a considerably larger capacity to receive information such as 2D or 3D images, although its temporal resolution is inferior to that of the tactile sense. Accordingly, to make the best use of the visual sense's ability to receive information, it is necessary for voice-based information expressed as temporal information to be indicated intuitively as spatial information. However, in converting voice information into 2-D images, there are some problems to be considered.

For example, when /ai/ is pronounced, listeners with normal hearing can hear the /a/ and /i/ separately, although the spectrograms flow continuously from /a/ to /i/. Stated more concretely, listeners do not perceive that a sound equivalent to /e/ exists

© Springer International Publishing AG 2017
T. Ifukube, *Sound-Based Assistive Technology*,
DOI 10.1007/978-3-319-47997-2_5

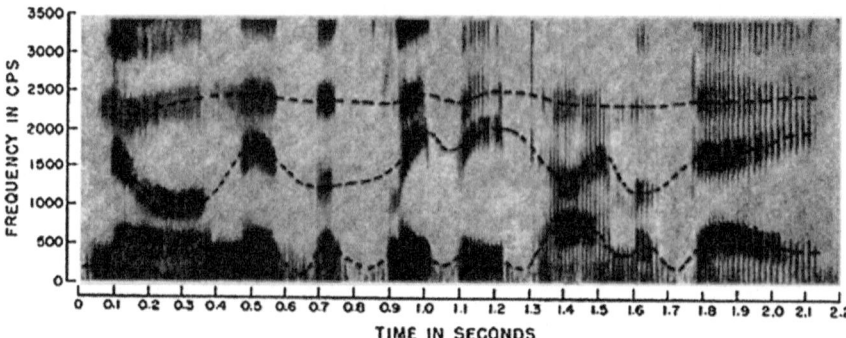

Fig. 5.1 Sound patterns in *visible speech*

between them. In a general sense, this phenomenon is referred to as "segmentation of the auditory system and/or the language cortex" as mentioned in Chap. 4. In short, the human brain has the ability to separate sounds discontinuously, even when the words are continuously spoken. For some time, researchers have debated the possibility of whether the visual sense may also possess a segmentation-like function. However, when the continuous sounds /a/ and /i/ are actually tested on a spectrogram, the /e/ spectrum can be seen. On the basis of this kind of analysis, whether a visual segmentation-like function exists or not remains unclear. As further evidence, most experiments conducted in the 1970s involving strict reading of spectrograms did not find adequate proof of the existence of any such visual segmentation.

In 1968, Liberman et al. [2] published a paper titled "Why are speech spectrograms difficult to read?". It should be pointed out that they claimed that it was essentially impossible for the visual sense to adequately read speech spectrograms, despite the use of highly processed images and extensive training, because the visual sensory system lacks a segmentation-like function comparable to that of the auditory sensory system.

Sometime later, however, a male specialist in linguistics showed the possibility of visual segmentation. The person in question had studied acoustic phonetics as well as English phonemic rules, and had been reading phonetic spectrograms consistently for more than 10 years (a total of 2000–2500 h). It was shown that he performed quite well in sentence-decoding experiments. Specifically, after considerable thought was put into the composition of 24 sentences of spectrograms, he attempted to decode them into a detailed phonemic series. When the results were examined to determine how closely his reading corresponded to the phonemic sequences, three phoneticians who acoustically evaluated them concluded that his correct answer rate was a surprising 85%. From such results, real-time systems were developed to enable the hearing impaired to better understand the correspondence between speech and spectrograms as well as to undergo vocalization training.

However, the question remained as to how well people with normal hearing would perform in such tasks. To investigate this, eight subjects were tested after having first been taught how to read spectrograms. More specifically, over a 20-h training period, they learnt spectrograms corresponding to monosyllabic voices spoken by one male speaker. The tests were conducted by using a list of 50 monosyllabic voices that were phonemically well balanced. The results showed that when the eight subjects read phonetic spectrograms presented in real time, their correct identification rate was better than 95%.

However, when they were presented with 50 completely new words without prior training, the rate of correct answers fell to only 60% for words, and 34% for the phonemics. From these visible speech experiments, in this research field, the dominant belief remained that the auditory sense is the only sense capable of perceiving continuous spectra separately, although research regarding visible speech was being conducted in different ways.

5.1.1 Speech Display Using Visual Color Illusion

On the other hand, in the early of the 1980s, Watanabe and his colleagues attempted to extract characteristic elements contained in each voice signal and described how the color-indication system of continuous speech can be utilized in vocalization training for the hearing impaired [3]. Thus far, such research has depended on visual illusions, especially *visual contrast effects*. For example, their research has indicated that by the use of visual characteristics corresponding to sounds such as /ai/, the visual contrast effects result from emphasizing the mid-part of the sound, thereby causing the /a/ and /i/ to appear as separate sounds.

The composition of this system is shown in Fig. 5.2. From the initial voice signals, the first, second, and third formant frequencies, the pitch frequencies and the distinguishing signals of the voiced and unvoiced sounds are extracted in real time. The pattern of continuous vocal sounds flows from the bottom to the top of the color TV screen, with each vocal sound usually appearing for about two seconds. Eventually, the pattern of continuous sounds appearing on the screen becomes a superimposed depiction of vowel phonemics that are mainly composed

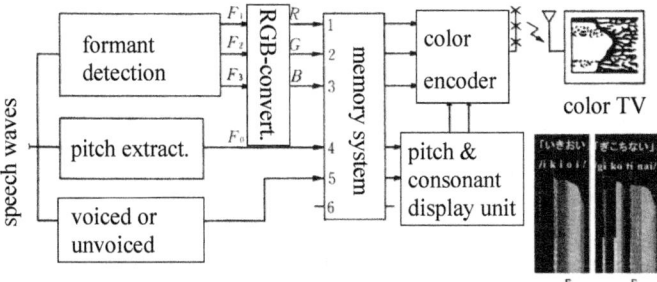

Fig. 5.2 Color display system representing continuous speech

of voiced sounds, intonations and unvoiced consonants. The unique characteristic of this visual pattern is that it makes use of color contrast—a visual illusion—to compensate for the changes in the formant frequencies caused by a phonemic environment. This effect of color contrast also clarifies the segment boarder of a continuous vowel.

As regards the recognition of continuous vowel sounds, an examination was conducted to determine the validity of the above-mentioned normalization of the individual differences between speakers with the color contrast effect. Sixty symmetrical three-vowel sequences uttered by two male adults and one female adult were randomly presented in real time and decoded by using this system. More specifically, in order to evaluate the compensating effect of acoustic sequences, the system attempted to decode the three-vowel sequences after having only learned single vowels. At the beginning of this test, although the correct answer rate was about 92%, it improved to 98% after only a few repetitions, even although three-vowel sequences had not been taught to the system at all.

However, it is still difficult to correctly answer the essential question of whether the visual sense can act as a viable substitute for the auditory sense to receive language-based information through images. Naturally, it will take many years of extensive research involving the hearing impaired to adequately answer this question.

5.1.2 Eyeglasses Displaying Speech Elements

Given that segmentation by the visual sense and the categorization of phonemes are difficult, some might consider showing them as separate patterns after signal processing. It goes without saying that although it is impossible to achieve perfect segmentation or categorization through signal processing, it can nonetheless be done with some accuracy. Upton's eyeglasses are famous for using this method [4].

As seen in Fig. 5.3, this principle shows characteristic information as virtual images beside the mouth of the speaker. It provides basic information regarding speeches as a supplement to lip reading. This information is very simple and is presented on a light emitting diode (LED) matrix that is composed of two quadrilaterals placed side by side, as shown in the figure. This tells us if the sound is a front vowel or a back vowel, as well as what kind of fricative or plosive consonants when any line of these quadrilaterals flashes on and off.

Special glasses for lip reading have also been developed to indicate characteristic speech information for peripheral vision. In speech training, it is becoming more important to provide information regarding articulation intuitively, rather than aiming to improve the recognition rate, as in the case of the normal recognition of speech. Thus, various methods have been proposed to use voice signals in order to display characteristic information for articulation, just as in Upton's eyeglasses. For example, in some studies, only the pitch frequency was extracted and presented as a

Fig. 5.3 Principle of Upton's eyeglasses

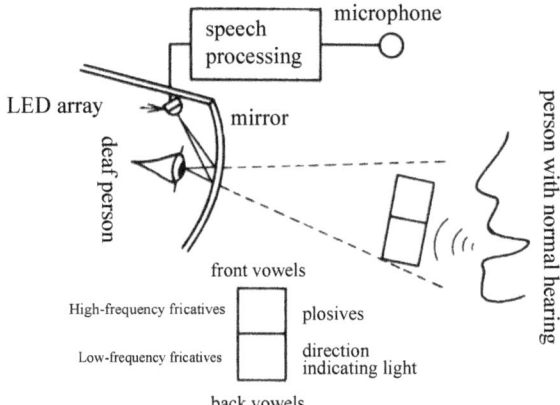

temporal pattern on a display, whereas other studies have involved the extraction of the $F1$ and $F2$ formants to present them in 2-D coordinates [5].

Furthermore, by combining virtual reality technology with speech processing, it is expected that new assistive devices for the hearing impaired will be invented. For example, simply by wearing a head mounted display (HMD) combined with a device that converts sound spectral patterns into images at a fixed distance in front of a speaker's face, the quantity of information available to the hearing impaired will be vastly increased. However, in this regard, basic research will be necessary to ascertain what sound quality is most easily combined with image quality of the visual sense. In any case, it is time to consider a method that can take full advantage of virtual reality technologies to assist the hearing impaired.

5.2 Application of Speech-Recognition Technology

The author and his colleagues once investigated the use of a tactile vocoder for people with acquired hearing disability. However, the subjects of our investigation expressed dissatisfaction with the device. It would seem that they expected to hear speech sounds inside their brain upon touching the vibrator arrays with a fingertip. As people with acquired hearing disabilities usually communicate by writing, they had high expectations of better alternative aids. In the case of acquired hearing disability, a much greater effort is necessary to learn the more traditional methods that the congenitally and severely hearing disabled can use, including sign language and lip reading. Accordingly, it would be highly beneficial for people with acquired hearing disabilities, if speakers used a voice-activated word processor that could serve as a substitute for sign language, lip reading and writing-based communication.

In fact, the greatest expectation for auditory substitution by means of the visual sense is the invention of a device capable of instantly transforming the spoken words of any speaker into written characters that can be easily read by the hearing impaired. This kind of device could also serve as a faster means for even the congenitally deaf to form language concepts. In order to realize this ultimate voice-recognition system, it will first be necessary to clarify the structure of the language cortex through quantitative analysis as well as to accumulate more extensive research findings on the structure of language itself from the viewpoint of engineering. In the field of voice recognition, problems cannot be solved solely by progress in signal-processing engineering, but must be coupled with significant advances in the fields of physiology and linguistics. Needless to say, the number of research engineers who have taken up the challenge of better understanding the language cortex is on the rise.

Furthermore, one should always bear in mind that such a voice-recognition system will mainly be effective for those with acquired hearing disabilities, as it is difficult for the congenitally deaf to form speech-based concepts by merely seeing letters that have been converted from speech.

5.2.1   Brief History of Speech-Recognition Systems

Research on voice recognition has been taking place since around 1950. However, progress has been limited by the fact that all of the apparatus necessary to conduct such research had to first be manufactured as hardware. Fortunately, in the 1960s it became possible to input speech waves directly into a computer or to input the sounds after first analyzing them by running them through an analogue filter bank. As a result of such progress combined with a basic understanding of the fundamentals involved, research focusing on speech gained acceptance as an academically approved field. The late 1960s witnessed the design of linear prediction coding (LPC) analysis (see next section) and saw the design of the dynamic planning (DP) matching method that regulates the elasticity of word duration by using a DP method.

These two methods became the mainstream for the technology of speech recognition throughout the 1970s. Various spectral comparative methods derived from parameters gained through linear prediction analysis in addition to some DP methods for recognizing speech were successively developed. Moreover, by using digital signal processing (DSP), it became possible to demonstrate various forms of speech recognition in laboratories.

In the 1980s, as it had become possible to deal with enormous amounts of data, orthodox methods for pattern recognition of speech through statistical analyses such as Bayes' theorem became the new standard. The Hidden Markov Model (HMM: see the next section) appeared within that framework. The HMM consists of DP matching and probability statistics to establish models for changes in voice patterns due to different speakers and contexts. Since that time, voice-recognition technology has been experiencing rapid progress.

Since the late 1980s, expectations that artificial neural networks would offer an innovative method for recognizing temporary voice patterns triggered their application to speech recognition. But despite the exceptions, this innovative method did not clearly succeeded in eliminating the need for conventional statistical pattern analysis. However, artificial neural networks have contributed to producing a new concept of artificial intelligence (AI) systems that have, recently, been applied to speech-recognition systems.

Sound-recognition technology has been developed and utilized in various ways to fit the purpose of utilization. However, the development of continuous speech-recognition technology that can recognize each spoken word—no matter who the speaker is—in normal conversation remains a challenge. But it may well be a helpful supporting tool for the hearing impaired, particularly for people with acquired hearing disability. The technology would also be useful to enable anyone to write a sentence without a keyboard and to communicate with computers and robots.

However, development of a device that can recognize speech information as effectively as a human being is almost beyond the capability of current technology. Discovering more about the speech-recognition function of the human brain would greatly contribute to the design of automatic speech-recognition systems. But at this moment, it is difficult to identify the brain function that analyzes, recognizes and understands speech information. Actually, it was impossible with the technology available a decade ago to manufacture a voice word processor that could convert natural and continuous speech into text. Despite the limited recognition accuracy in today's technology, it is still fair to say that speech-recognition devices are helpful for the hearing impaired as inferential abilities can fill in the gaps of the device.

As many hearing-disabled people have strongly requested such a device, the author and his colleagues began research on developing a voice-activated word processor (named the Voice Typewriter) that was capable of at least receiving non-continuous spoken words through a microphone and processing those sounds into text by using a microcomputer. The author would now like to offer a brief description of the Voice Typewriter that was designed in 1977.

5.2.2   Monosyllabic Voice-Activated Word Processor [6]

The pre-processing stage of our Voice Typewriter adopted the general methods of the speech-recognition system shown in Fig. 5.4. The process that this figure illustrates is that by imitating the properties of the outer ears, high-frequency speech is emphasized and received through a microphone at 6 dB/oct. These characteristics are then extracted by the same process used by the inner ears. The auditory sense does not only process voice signals on the temporal axis, but also converts them on the frequency domain and conveys spectrum envelop information to the CNS. Although there are various ways to detect spectrum envelopes we designed a

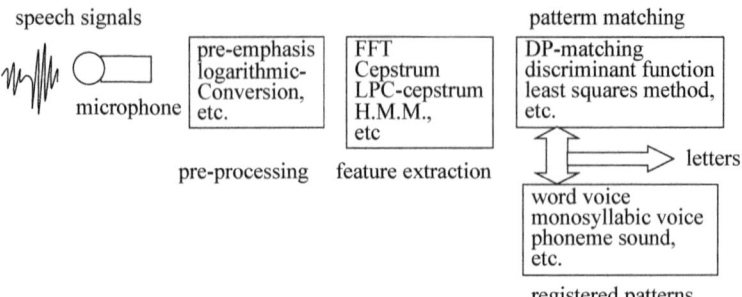

Fig. 5.4 General structure of automatic speech-recognition system

band-pass filter bank that is modeled after the "critical band" assumed in the human auditory system.

Fortunately, Japanese speech is represented as a series of 68 monosyllables (/a/, /ka/, /sa/, /ta/, /na/......), each of which consists of five vowels and 14 consonants and is represented by a Japanese *kana*. This simplicity made it easy to design the Voice Typewriter, which converts each monosyllable into a corresponding *kana*.

Thus, as Fig. 5.5 shows, our first method was to utilize the output of a tactile vocoder's filter-bank via an A-D converter and, by using a microcomputer, recognize this monosyllabic speech. As for pattern matching, our method was to first distinguish the five vowel groupings of the Japanese *kana* system of syllabic writing from one another. After determining into which *kana* grouping a monosyllable fell, the computer further distinguished which unique sound it was. For example, if the monosyllable happened to be /sa/, it would fall within the vowel grouping containing the monosyllables /a/, /ka/, /sa/, /ta/, /na/, /ha/, /ma/, /ya/, /ra/, and /wa/. Finally, the result was indicated as a *kana* character typed on a computer screen. The time spectrum pattern of the 68 monosyllabic voices spoken by the unique voice-pattern of each speaker was pre-registered in the computer's memory.

Fig. 5.5 Block-diagram of a Japanese monosyllabic voice-activated word processor (the Voice Typewriter)

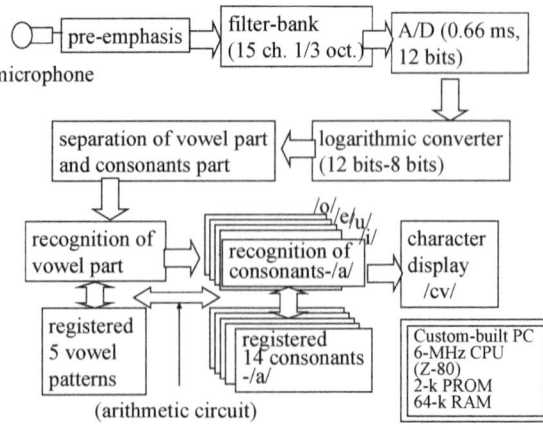

Fig. 5.6 Appearance of the Voice Typewriter using a Z80-microprocessor and 32-kbyte memory

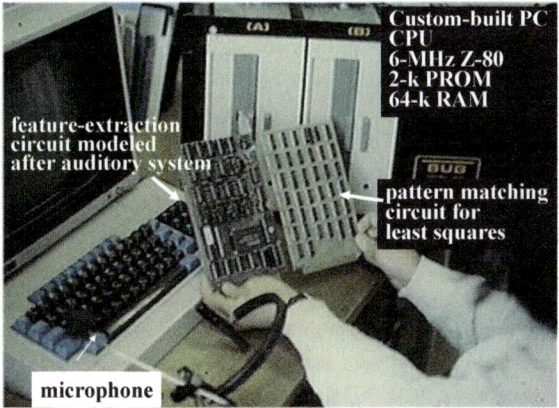

At the middle of the 1970s, a CPU known as the Z80, was developed in the United States. We used this CPU to manufacture a micro-computer with 32 k-word memory (RAM with 64-kbyte memory) that was connected to a voice feature extraction circuit as well as a voice pattern matching circuit (Fig. 5.6). By this method, it became possible to determine the most suitable parameters for speech recognition and to eventually raise the recognition rate to about 96%.

Table 5.1 shows the confusion matrix for the recognition of 68 monosyllables for five speakers. Although the misrecognition rate for /p/, /t/, /k/ was noticeably high, there were no other significant problems. Although this device enabled

Table 5.1 Confusion matrix for consonant recognition using the Japanese monosyllabic Voice Typewriter

average of 5 speakers

↘ out in ↘	(v)	k	s	t	n	h	m	y	r	w	g	z	d	b	p	¥ [%]
(v)	493					1	1				4				1	98.6
k		460		8	28						4					92.0
s			2	479	14							5				95.8
t			5	27	463							1			4	92.6
n					492		5	2					1			98.4
h			10	1		487							1	1		97.4
m					21		479									95.8
y								300								100.0
r					3	1	2		489		1	1		3		97.8
w										99				1		99.0
g		2	1			2	1				487	5		2		97.4
z			1	1	4						2	492				98.4
d					1							1	295	3		98.3
b		1								1	2	1	2	490	3	98.0
p		5	1		13	2	8							17	454	90.8
												total ave.				96.4

hearing individuals to input words into a computer, its initial development was primarily to aid people with acquired hearing disabilities.

It is necessary to pay attention to the peculiar characteristics of Japanese language when converting vocal sounds into characters. Japanese characters consist not only of the "phonogram" that is understood upon its meaning as a sound, but also the "ideograph", namely *kanji* (Chinese characters), whose meaning is understood upon its visualization. Furthermore, an error in *kanji* conversion might lead to a total misunderstanding of a sentence as several homonyms exists for each *kanji*; that is to say, many *kanji* share the same pronunciation even although their meanings are completely different. Fortunately, as Japanese word processors that can automatically convert *kana* series into *kanji* with a certain degree of accuracy were developed and manufactured in the 1970s, any spoken sentence could be converted into *kanji* together with *kana* by connecting the Voice Typewriter to a word processor.

It goes without saying that voice-activated word processors are not effective for every hearing-impaired person because it is not easy for congenitally deaf people to extract meaning and concepts—even when a voice-based language has been converted into text. However, as a substitute for written communication, we developed a device that is held in a speaker's hand and into which he/she speaks words that a hearing-impaired person can see displayed on LCD eyeglasses, as shown in Fig. 5.7.

This could be used at home or even to assist a speaker who is delivering a lecture to many hearing-disabled people. Considering that this could be used to input information into a computer for people suffering from a disability that inhibits natural hand movements, it is conceivable that such an input system could be developed as a pocket-size portable device. In view of today's considerable advancements in technology, it is certainly worth taking a second look at the feasibility of such an endeavor.

Fig. 5.7 Eyeglass display presenting characters while showing the speaker's face and lip movement

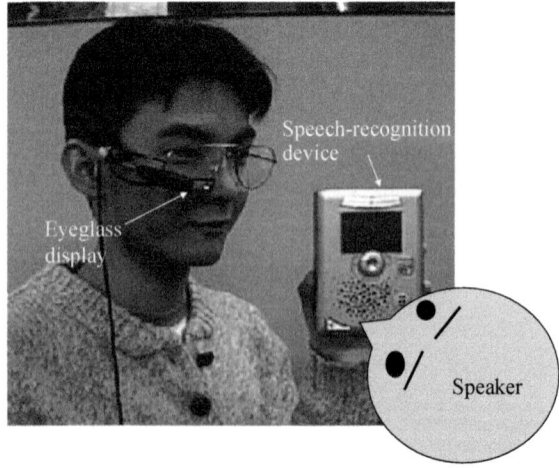

However, in the period around 1980, the high cost (around $20,000) of this device made it too expensive to purchase for most private individuals. It was used for some time by printing companies to input spoken words into word processors and could convert recognized *kana* into mixed *kana-kanji* text.

5.2.3 Basic Signal Processing for Recent Speech-Recognition Systems [7]

As a great number of books and papers have been published on voice-recognition systems, the author will limit this discussion to the concepts below. Recent progress in speech-recognition systems has been greatly dependent on signal-processing algorithms as well as computer hardware. As described in the next chapter on vocal cord models (see Chap. 6), the acoustically conveyed function of vocal chords can be replaced by a digital filter given that speech waves are reproduced by a delayed linear input of impulses.

Digital filters are generally based on linear polynomials (see block diagram in Fig. 5.8), which that is a basic concept of LPC. Furthermore, the LPC spectral envelope has become mainstream and has replaced analog filter banks. The coefficients a_1, a_2, a_p, etc., are referred to as prediction coefficients, and linear prediction is the method used to predict the next time value. When the conveyed function for this system is indicated by z-conversion, it becomes a multipolar model that solely consists of poles, as shown by (5.1).

$$H(z) = 1/(1 + a_1 z^{-1} + \cdots + a_p z^{-p}) \tag{5.1}$$

Without delving into the details, z^{-1} indicates the delay element in sample time (*T*) and can be indicated as a frequency function. When we calculate $|H|^2$ by specifying z^{-1} as the frequency and substitute it into (5.1), the power spectrum for the optional frequency (*f*) can be obtained. As the power spectrum obtained by the LPC corresponds to the spectral envelop, which does not contain information

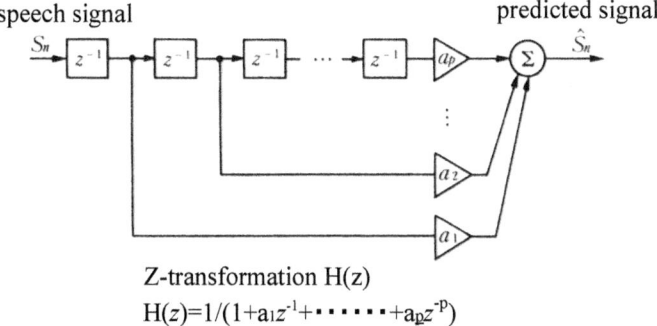

Fig. 5.8 Block diagram showing the linear prediction

Fig. 5.9 An example of an
LPC spectral envelope

vowel: /a/ , frame length: 12.8 ms,
Hanning window, LPC order: 21

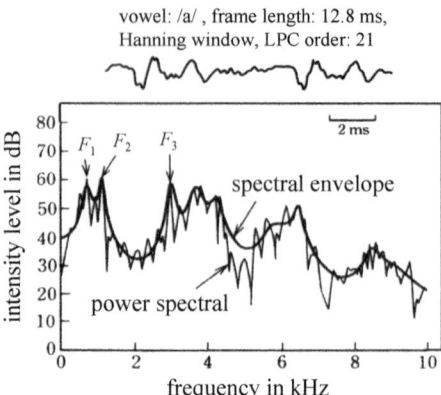

regarding the sound source, some peaks correspond to the formant of the spectrum, as shown in Fig. 5.9. When obtained by this method, the spectral envelop is often used as a characteristic parameter.

When the power p is twice the number of formants, the peak of the LPC spectral envelope corresponds precisely to the formant. For a 16-kHz sampling frequency, the most appropriate value of p is about 20. In the case of a female voice, which has a high resonance, a value of 22 is more suitable.

Basically, a time spectral pattern obtained by the LPC method is divided into several equal components. When n components on the frequency axis are divided into m on the temporal axis, an $m \times n$ matrix is formed. As an example, one word can be expressed as a single point on the $m \times n$ spatial-dimensional matrix. When a certain point on the matrix moves in any direction, this traced trajectory corresponds to one of the sentences. Within this matrix, a discriminate plane is created that classifies the patterns according to which side of the plane it is located.

The most commonly used method in voice pattern matching is to measure the distance between the matrix data $(m \times n)$ obtained from both the input speech signals and the registered speech signals in a computer memory, and to then search the registered speech for the one with the least difference. The more important problem is how to match the temporal axis; that is to say, how to normalize the time series of input speech data with standard speech data. Sound recognition in word units has been greatly improved since the invention in the late 1970s of linear prediction coding as a speech-production model, and the DP matching method that can normalize the different durations of each spoken word, as mentioned above.

5.2.4 HMM Algorithms [8]

In the 1980s, technologies became available that made it easy to handle large quantities of speech data. As a result, pattern-recognition methods that processes vocal sounds by statistical algorithms became mainstream. Among them, the HMM was developed, which dramatically improves sound-recognition technology. The HMM has been widely utilized in such environments as speech-captioning systems for TV broadcasting and in cellular phones. Since then, several improvements have been made in sound-recognition technologies. For example, big data of both spoken words and sentences accumulated from mobile phone usages have been added in the computer memory, whilst internet search technology—and recent AI algorithms that are used for pattern recognition—have been applied to the speech recognition software.

An example of a HMM algorithm is schematically shown in Fig. 5.10. In general, 13 LPC coefficients are used, which reflect the characteristics of the spectral envelope and also compress the speech information. Actually, LPC coefficients are estimated from an observation sequence of 20–30 ms extracted from vocal sounds at intervals of 10–20 ms.

The spectrum envelope in a short observation duration, say 30 ms, can be considered to be a steady state without time change. Moreover, a series of sounds can be assumed, namely in the Markov process. That is to say, the relation between a state, $(S_{n,t})$ at the present time (t) and a state, $(S_{n,t-1})$ at a previous time (t−1) are considered to be stochastically determined. Where, $(S_{n,t})$ is a candidate of observation sequence(O_t), and $(S_{n,t-1})$ is a candidate of the previous observation sequence (O_{t-1}). Actually, the Markov process is presumed to be between $(S_{n,t-1})$ and $(S_{n,t})$ and also between (O_{t-1}) and (O_t), which are extracted at intervals of 30 ms because the vocal sound does not rapidly change within the interval under the restriction of the articulation organ function. In this way, a speech-recognition method in which a sequence of vocal sounds is assumed to be a Markov process is called "HMM speech recognition," which presently is the most widely used method. However, in this method, it is necessary for a probability (state transition

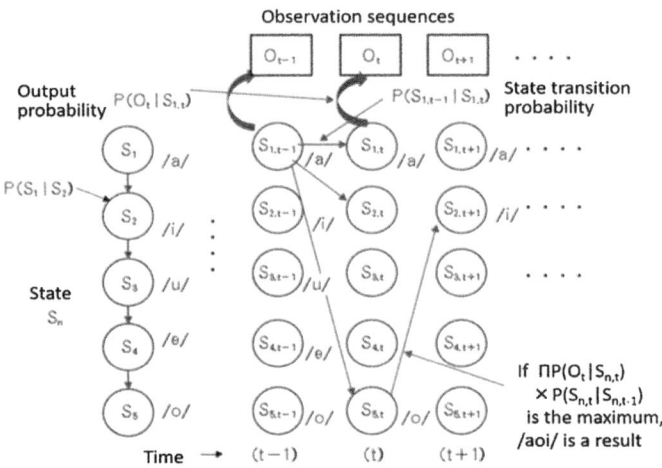

Fig. 5.10 Schematic illustration of HMM sound-recognition algorithm

probability) $P(S_{n,t}|S_{n,t-1})$, which moves to the present state $S_{n,t}$ from the previous state $S_{n,t-1}$, to be set in advance from among a large set of data.

In HMM algorithms, first the state S_n (i.e., a certain phoneme) at a time t is presumed to be $S_{n,t}$, then the probability $P(S_{n,t}|O_t)$ (output probability) is calculated in order to estimate S_n that may correspond to observation sequence O_t (in general expressed by LPC coefficients) at time t. It is most likely that the state with high output probability is the state S_n (i.e. a certain phoneme) at time t. However, if the transition probability $P(S_{n,t}|S_{n,t-1})$ is lower than the others, the recognition result is less likely to be S_n. The HMM is a method in which the relatively larger product $P(O_t|S_{n,t}) \times P(S_{n,t}|S_{n,t-1})$ is regarded as a candidate of recognition results. To facilitate the understanding on the HMM algorithm, assume that the state S_n is one of five Japanese vowels /a/, /i/, /u/, /e/, /o/ as shown in Fig. 5.10.

In the figure, the observation sequence is O_t, and state S_n is one of the five Japanese vowels /a/, /i/, /u/, /e/ and /o/. If the result for a certain state is S_n (/a/, /i/, /u/, /e/, /o/), the probability that S_n is an observation sequence O_t can be represented by five output probabilities $P(O_t | /a/)$, $P(O_t) /i/)$, $P(O_t | /u/)$, $P(O_t | /e/)$ and $P(O_t | /o/)$. The observation sequence O_t with the highest probability among the five output probabilities is mostly like to be S_n that may be a recognized vowel at time t. However, if the transition probability $P(S_{n,t}|S_{n,t-1})$ was low, O_t might not be S_n. Therefore, it is necessary to previously determine the probability $P(S_{n,t}|S_{n,t-1})$ from a database and to calculate the products of $P(S_{n,t}|S_{n,t-1})$ by $P(O_t|/v/)$ (v:/a/,/i/,/u/,/e/, /o/). In this case, the products are 25 in total. $S_{n,t}$ is regarded as a recognized vowel whose product shows maximum among the 25 products.

If the vocal sound fragment is one phoneme, its recognition result can be identified as a certain phoneme. However, in general, the phonemes are connected with many sound fragments. Therefore, the maximum product should be determined among all products $[P(S_{n,1}|S_{n,0}) \cdot P(O_1|/v/)] \cdot [P(S_{n,2}|S_{n,1}) \cdot P(O_2|/v/)] \cdot [P(S_{n,3}| S_{n,2}) \cdot P(O_3|/v/)] \cdots$, where /v/ is one of /a/, /i/, /u/, /e/ and /o/. The traced phonemes whose product shows a maximum may be recognized as a phoneme series. For example, /aoi/ is a recognized phoneme series that is /a/→/o/→/i/ in the case shown in the figure.

To simplify the above calculation, all state transition probabilities (for example, the probability that transits from vowel /a/ to vowel /i/ [P(/i/|/a/)]) should be determined by a 'phoneme model' that has been estimated by a large number of phonetic data beforehand. In addition, to further simplify the calculation, the word transition probability (for example, a probability that transits from the word "you" to the word "are," [P("are"|"you")]), which is called a 'language model', should been also previously determined from a large number of sentence data. Given that the more phoneme data and language data there are, the more precise the phoneme model and the language model are, the more the recognition precision can be improved.

The recognition precision can be improved by creating models such as "sentence structure," "meaning," "context," "scene," and "feelings"—as is the case with the phoneme and the language models. Actually, the correct recognition rate has been greatly improved by utilizing the enormous amount of sentence data collected on the internet, the so-called "big data" that has become available for use in recent years. One piece of software that has benefitted from this big data is "Siri," which Apple, Inc. has provided as an application software for smartphones.

In addition, natural language processing (NLP) by AI can be applied for further improvement of speech recognition. AI such as "Deep Learning" is basically a computer algorithm modeled after the human brain mechanism. In recent years, the AI installed in robots can even engage in a natural conversation with human beings. Scientific endeavors have been conducted to utilize AI robots as a substitute for communication partners as well as a memory supporting tool for those with mild cognitive impairments (MCI). This topic will be further elaborated at the end of this chapter.

5.3 Captioning Systems [9]

Despite the rapid progress in speech-recognition technology, it remains difficult to understand the context automatically, and the user of speech-recognition equipment needs to select the most suitable word from among similar words having the same pronunciation. Another problem with current speech-recognition technology is that the recognition rate decreases immediately after the speaker or the topic changes. In addition, speech-recognition equipment must be used in quiet environment with a special microphone held in a predetermined position near the mouth of the speaker.

When a grammatical construction such as "It was a cold day tomorrow" is made, the sentence is considered to contain a contradiction as the word "tomorrow" indicates the future, whereas the term "was" suggests the past. As such, the above sentence is judged to be incorrect according to rules of grammar. In addition, human beings in general scrutinize information and guess intended meanings by listening to speech sounds. During this comprehension process, they rely on words that they already know in order to judge the accuracy of their interpretation.

In fact, people unconsciously reprocess any linguistic information that has been misunderstood or appears ungrammatical by taking advantage of the overall context: that is to say, they use a top-down approach. Furthermore, to reconfirm the correct interpretation of a vocal sound, for example, if they mistakenly hear the letter /p/ when the speaker actually meant the letter /k/, listeners repeat the pattern-matching process once again. In summary, they recognize what a speaker means by alternating between the use of both bottom-up and top-down approaches in order to ensure correct understanding. It is important to make good use of this human ability in the captioning system for the hearing impaired that is to be explained next.

The author and his research associates designed a captioning system that could automatically convert various languages into both Japanese and English for the 6th Disabled Peoples' International (DPI) conference held in Sapporo, Japan in 2002, which was attended by around 3000 disabled people from more than 100 countries. As the speech-recognition technology used was not perfect, we expected that this technology would need to be combined with the ability of the hearing impaired whereby they can often guess the correct meaning of spoken sentences by observing a speaker's face and lip movements.

With this in mind, we designed the captioning system in such a way that both the series of characters, including *kanji*, and the speaker's face simultaneously appeared on a large screen. We used a commercially available speech-recognition program called Via Voice made by IBM Corporation. Furthermore, we adopted the "shadowing speaker" method in which the speaker's words are sent to a well-trained shadowing speaker who then repeats them. This is almost the same method as the "re-speak method" that NHK (Japan Broadcasting Corporation) proposed around 2000.

In our system shown in the upper figure of Fig. 5.11, as the recognized outputs were checked by human beings, Japanese and English characters with a better-than 98% correct recognition rate could be displayed on the screen. The captioning system was first used at a preliminary conference held in 2001 in order to check the system. At the DPI conference in 2002, both Japanese and international sign languages, Japanese and English captions, as well as speakers' faces were displayed on a same screen as shown in the lower figure of Fig. 5.11.

Based on the analysis of the captioning results, it was found that captioning speed greatly depends on the language: Japanese or English. In order for the correct recognition of the captioning to reach about 98%, the captioning speeds were 4 s from English to English, 11 s from Japanese to Japanese, 12 s from Japanese to English, and 17 s from English to Japanese, as shown in Table 5.2.

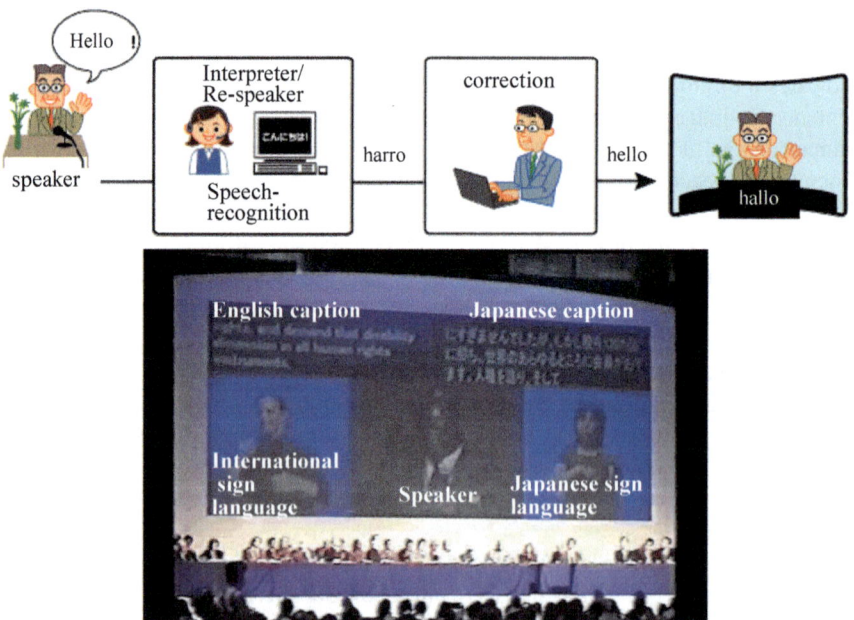

Fig. 5.11 Captions (English and Japanese) presented on a screen together with sign language and the speaker's face at the DPI conference in 2002

Table 5.2 Correct recognition rate and captioning speeds

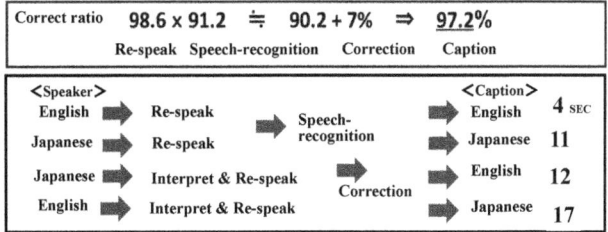

After using the system at the DPI conference, we investigated how correctly nonverbal information such as a speaker's lip movements and facial expressions (shown in the left of Fig. 5.12) could improve the comprehension of spoken sentences that contain incorrect words [10]. As shown in the right of the figure, although the results were different among the deaf subjects, the following tendencies were clarified. That is, the highest rate of the correct recognition was obtained when the lip movement was displayed with captions, and a higher rate was obtained when facial expressions were displayed with captions than when only captions were displayed. These improvements were observed when both the lip movement and the facial expressions were displayed roughly one second after the

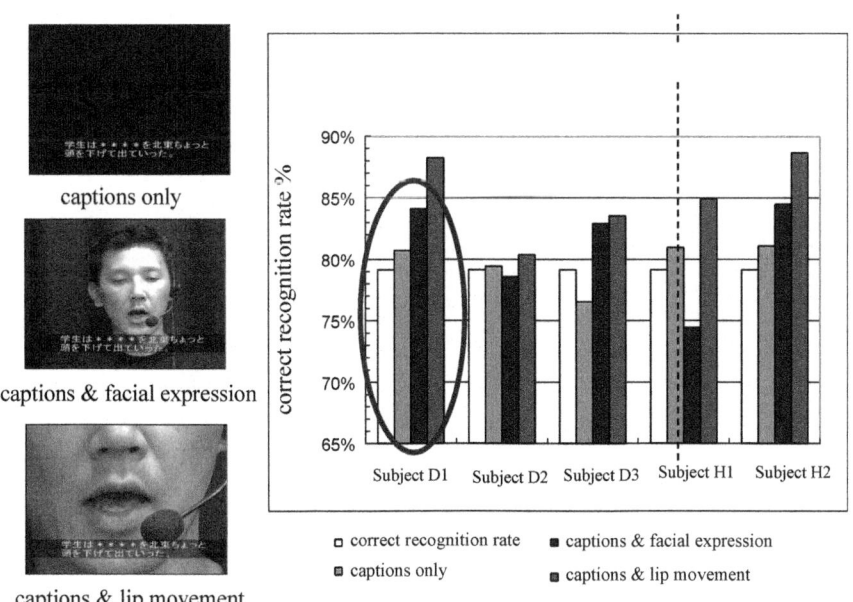

Fig. 5.12 *Top* Caption only (*left*), caption + face (*center*), caption + lip movement (*right*), *bottom* recognition rates

incomplete captions. No improvement was obtained for people with normal hearing. It was ascertained that combining both incomplete verbal information with nonverbal information is indeed significant in facilitating comprehension for the hearing impaired.

Furthermore, this is ascertained from another experimental result that was obtained by a deaf researcher NAKANO and her colleagues [11]. Their results show that the eyes of congenial deaf subjects (eight persons aged 25 to 37) are not staying on the captions as if they are looking for some clues besides the captions, differing from the subjects with normal hearing as shown in Fig. 5.13. Therefore, we designed a new captioning system that can display the speaker's facial expressions and lip movement on a screen after a certain delay, together with the corresponding text. After trying the captioning system at the DPI conference, it was used at various conferences in Japan. For example, at the International Conference of Universal Design held in a conference room in Yokohama City (see Fig. 5.14) in 2002, both English and Japanese speech picked up in the conference room were simultaneously interpreted, and the interpreted speech was sent to place A (Sapporo), in the figure, where the speech was translated into both Japanese and English texts that were displayed on screens in the conference room.

The author and his colleagues attempted to apply the captioning system to a communications aid for mobile phones in 2004 for use by people without disabilities. To offer another very plausible application, we can envision the creation of a device for the hearing impaired consisting of a liquid crystal display (LCD) on one side and a microphone on the other side, into which a speaker's words will instantly be projected onto the opposing screen as characters. In the near future, people will be able to carry such electronic devices in the form of a wristwatch. These experiences suggested that assistive tools should be as widely used as possible, that is, not only for deaf users but also for non-deaf users, in order to make the system better and more affordable.

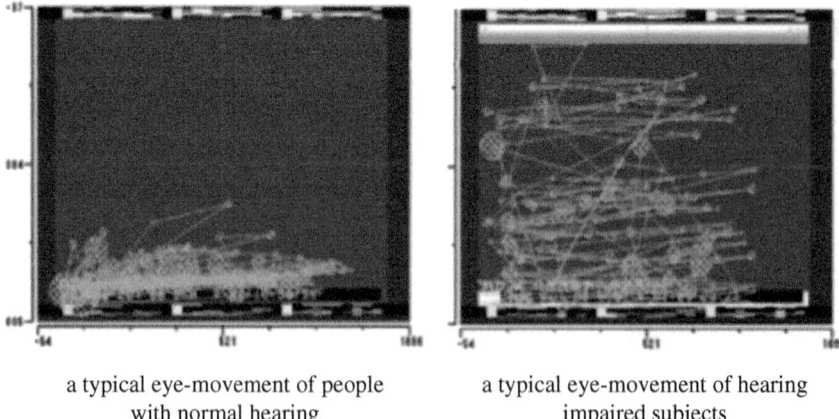

a typical eye-movement of people   a typical eye-movement of hearing
with normal hearing   impaired subjects

Fig. 5.13 Eye movement of hearing impaired-people and people with normal hearing while looking at a caption scrolling from bottom to upper on a screen

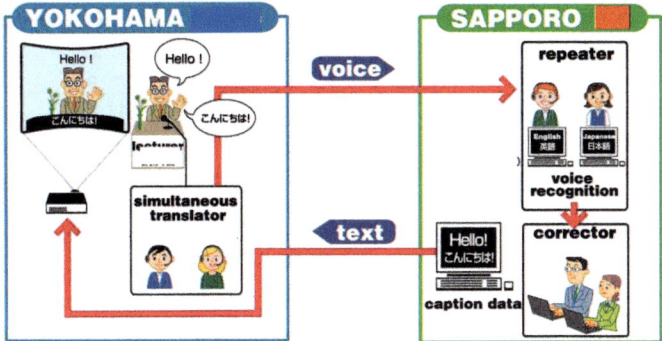

Fig. 5.14 Block diagram of captioning network system

The captioning system is useful for interpreting at international conferences where different languages are used. By using the internet, the system will be available to a user wherever he or she is, by transferring the speech and images of the speaker to an interpreter or a re-speaking person working at home or at a remote place. The captioning system also has the potential to allow a repeating person and an interpreter to conduct home-based business and an impaired person who finds it difficult to leave their home to work as a repeating person at home.

5.4   Communication Robots for Cognitive Declined People

A project of the Life Support Robot Systems that has been carried out by joint research conducted by the National Rehabilitation Center led by Inoue and NEC Ltd. under the author's supervision of JST project mentioned in Chap. 1. It aims to support independence of elderly persons and to promote their social participation.

5.4.1   Cognitive Declined People in a Super-Aged Society

This project targets the ever-increasing population of elderly people with dementia (cognitive impairment) and those with the potential risk of developing dementia who have MCI (mild cognitive impairment) and/or memory decline in a Japan that is becoming a super-aged society. According to a report published by the Ministry of Labour, Health and Welfare in 2010, it is estimated that the total number of people in Japan with dementia is around 3.45 million, and those with MCI is about 8.37 million, as shown in Fig. 5.15 [12]. Those numbers are forecasted to continue to increase to near 10 million (around 10% of the population) in 2025. As the potential risk for developing dementia becomes higher after 75 year of age, social participation could be highly significant in delaying the "at-risk age" in the elderly.

Fig. 5.15 Number of people with MCI and dementia in Japan (2010) [12]

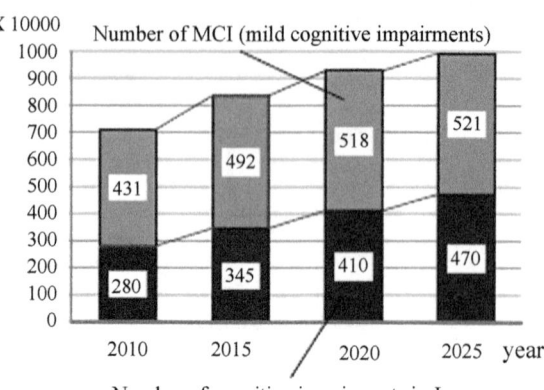

Number of cognitive impairments in Japan

5.4.2   Design Themes for Communications Robots

In preliminary research, it was found that a robot is the most suitable form of communication with the elderly from among other tools—such as TVs, smartphones or tablet PCs—as it most favored by them; positive answers such as "likable" or "rather likable" for a communications method using robots reached 55%, whereas TVs and tablet PCs received 37.7 and 37.9%, respectively. These results show the potential use of robots for communicating with the elderly.

In this project, optimal robot systems, as shown in Fig. 5.16, have been constructed that consist of two aid functions for the elderly activated by interactive conversation with the robot. The first is "memory assistance", designed to support users who have MCI and/or memory decline by helping them manage their daily schedules inside their nursing homes or similar facilities for the elderly. The second is "action support", intended to guide the actions of the elderly outside their nursing homes. The elderly can communicate with the robot using a spoken diagram that is designed and customized for each individual by using the robot's ears and mouth and also behavior observation by the robot's eyes.

5.4.2.1   Needs of the MCI [13, 14]

Optimal robot systems will extend the role of life-support robots to total systems including wearable phones and tablet computers that enable communication even outside nursing homes and the transfer of information to hospitals, drugstores, etc., as shown in the right side of Fig. 5.16. The robot systems can also function as an intermediate for the elderly inside nursing homes and their families or outside health management facilities to transmit information and feedback between the two parties.

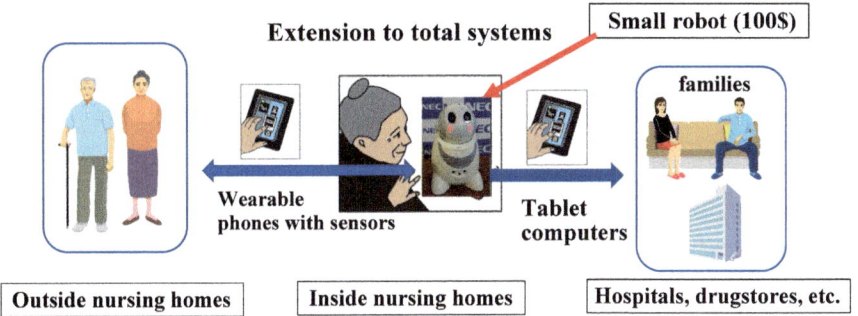

Fig. 5.16 Life-support robot systems

As a result of a survey conducted among 124 persons including the elderly, families of the elderly, care managers and device developers regarding the usage of robot systems, 172 needs were ascertained including management of daily schedules and taking medication, support for blood-pressure checks, calorie calculation, and the need for a referee on conflicts among the elderly that might result from their dysfunctioning memories. Based on the results, 36 kinds of spoken diagram were designed with the aim of complementing the memory function of the elderly.

However, the communication robot requires a good deal of improvement so as to be able to understand the obscure speech patterns of elderly people with dementia and MCI. Moreover, another function based on the structure of conversation is necessary in order to check whether the information from the robot is successfully understood by the elderly. It is assumed that the elderly might fail to understand the verbal communication by the robot caused by their declining abilities in listening and memorizing. To tackle such miscommunication between the robot and the elderly, a spoken diagram technology with speech synthesis and recognition method based on structure of conversation has been designed.

5.4.2.2   Conversation with Robots

The spoken diagram with a robot is divided into four elements; (a) Alert and reminder, (b) Preceding, (c) Information transmission, (d) End of conversation, as shown in Fig. 5.17a. For example, the robot starts the conversation at the appointed hour, asking "Have you taken your medicine?" Then the robot recognizes the response from the elderly concerned as "Yes, I have." The conversation ends when the robot answering "Very good" in response to the verbal communication from the elderly indicating the medicine has been taken.

Inoue and his co-colleagues conducted a survey of 20 elderly persons with dementia and MCI to check the accuracy in communication using robot systems

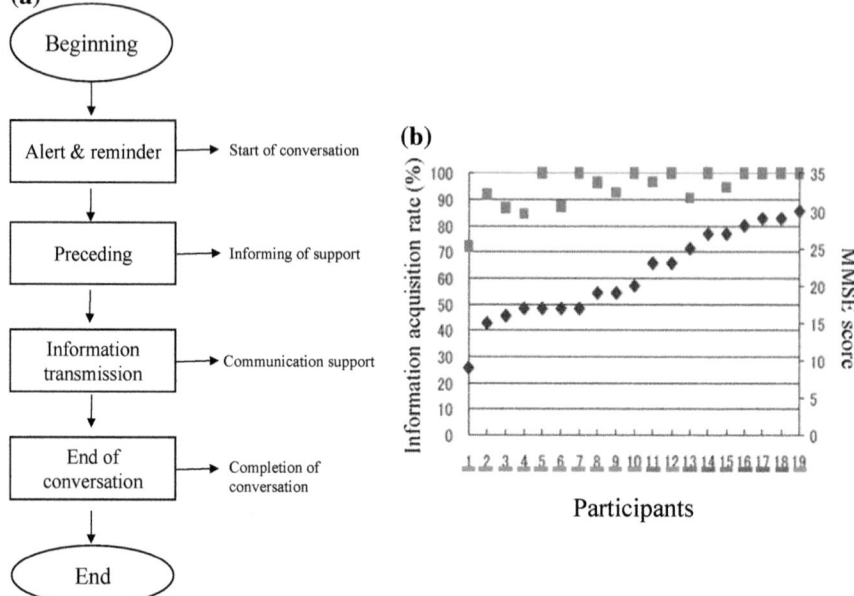

Fig. 5.17 a Interactive conversation between a robot and an elderly person with MCI, **b** information acquisition rate/MMSE score

that implement the spoken diagrams. The result showed that the information acquisition rates among the elderly—irrespective of the degree of dementia, as ascertained by a mini-mental state examination (MMSE)—were better than 80% in communication with the robot, as shown in Fig. 5.17b. On the other hand, the survey result for five elderly persons who lived alone showed almost 100% responses for alerts and reminders from the robot, and around 90% correct diagrams were achieved for correctly interpreting information transmitted by the robot. From these evaluation tests, it is anticipated that speech technology can play a key role in the support of the elderly with dementia and MCI by being used in robot systems.

5.5 The Future of Visual-Based Auditory Substitutes

Although speech-recognition technology has various goals, depending on the intended use, the most difficult hurdle to overcome is enabling such a device to automatically recognize the continuous speech of any number of speakers whose individual speech patterns and pronunciation naturally differ to some degree. If this goal is realized, it will undoubtedly prove to be of the greatest benefit for the hearing impaired, especially for those who have acquired hearing disabilities.

Furthermore, it could be used as a means of input for word processing by those who have lost the function of an upper limb or a cognitive function.

As described in Chap. 7, we can soon expect great strides to be made in virtual reality technology. If we are able to receive sound-related information through such HMDs or eyeglass displays as peripheral vision, not only will we be able to receive voice-based information, but also other sound-based information related to our environment, such as warning sounds. If that is the case, research will be necessary to determine specific factors such as what kinds of sounds should be converted into what kinds of images. As such matters are closely related to the field of unknown sensory and brain functions, it is indeed encouraging to know that many of those researchers are increasing their efforts to make progress on auditory substitution by visual sense.

References

1. R.K. Potter, G.A. Kopp, H.C. Green, *Visible Speech* (Van Nostrand, New York, 1947)
2. A.M. Liberman, F.S. Cooper, D.P. Shankweiler, Why are speech spectrograms hard to read? Am. Ann. Deaf **113**, 127–133 (1968)
3. A. Watanabe, Y. Ueda, Trans. Inst. Electron. Commun. Eng. **J-64**(A), 574–581 (1981)
4. H.W. Upton, Visual speechreader design considerations. in *Preprints of the Research Conference On Speech Processing Aids for the Deaf*, Gallaudet College (1977)
5. D. Ebrahimi, H. Kunov, Peripheral vision lipreading aid. IEEE Trans. BME **38**(10), 944–951 (1991)
6. Y. Nitadori, T. Ifukube, C. Yoshimoto, A real-time recognition system of monosyllabic voices. J. Acoust. Soc. Jpn. **39**(2), 75–81 (1983)
7. J.D. Markel, A.H. Gray Jr., *Linear prediction of speech* (Springer-Verlag, Berlin, Heidelberg, New York, 1979)
8. K.F. Lee, Hidden Markov models: past, present, future. in *Proceedings of Eurospeech*, 148–155 (1989)
9. S. Nakano, T. Kanazawa, T. Makihara, H. Kuroki, K. Ueda, S. Ino, T. Ifukube, A study on the ease of real-time speech-to-caption system for the hearing impaired (1): the effect of line break. Trans. Hum. Interface Soc. **10**(4), 435–444 (2008). (in Japanese)
10. H. Kuroki, S. Ino, S. Nakano, K. Hori, T. Ifukube, A display method of non-verbal information for an automatic captioning system used for the hearing-impaired. Hum. Interface (2004) (in Japanese)
11. S. Fukushima, S. Nakano, T. Kanazawa, H. Kuroki, S. Ino, T. Ifukube, Some issues to study developing and practicing the Real-time captioning system using automatic speech recognition technology. Academic Knowledge Archives of Gunma Institutes, Gunma University. **55**, 179–186 (2006) [in Japanese]
12. Comprehensive survey of living condition published by the Ministry of Labour, Health and Welfare in 2010
13. T. Inoue, R. Ishiwata, R. Suzuki, T. Narita, M. Kamata, S. Motoki, M. Yaoita, Development by a field-based method of a daily-plan indicator for person with dimentia. in *Assistive Technology from Adapted Equipment to Inclusive Environments*, (AAATE, Florence, 2009), pp. 364–368
14. T. Inoue, Field-based development of an information support robot for persons with dementia. Technol. Disabil. **24**(4), 263–271 (2012)

Chapter 6
Assistive Tool Design for Speech Production Disorders

Abstract In this chapter, the author discusses an artificial electro-larynx that can produce intonation and fluctuation of the larynx voice for speech disorders, especially for laryngectomees. In the design of the artificial larynx, he emphasizes that many hints were taken from the vocalization mechanism of a talking bird, the mynah. The author also introduces a voice synthesizer called "Let's talk by a finger" for people with articulation disorders that make it hard to control their voice organs because of neuromuscular disease or speech apraxia. The synthesizer, which can produce any speech sound just by touching and stroking the touchpad of a mobile phone, was modeled after the vocalization mechanism of a ventriloquist who can produce any speech sound without moving his lips. Furthermore, evaluation methods are indicated for aids and treatment of speech organ disorders caused by a cleft palate and a reversed occlusion.

6.1 Substitution of Speech Production

As described in Chap. 1, the speech organs of human beings are divided into three parts: the lungs to send expired air to the larynx; the vocal folds to convert the exhaled air into a sound source; and the vocal tract that extends from the larynx to the lips (right part of Fig. 6.1). In contrast, the primary functions of an animal's organs that correspond most closely to the vocal organs of humans are: the lungs for breathing; the larynx to protect the lower airway from foreign objects; and the oral cavity, larynx and nose to chew, taste, swallow, and for performing the function of ventilation. By using the same organs, human beings, however, have acquired a unique ability to produce vocal sounds, hence language.

The main reason for performing a laryngectomy is cancer of the larynx. However, one of the main causes of this cancer is believed to the fact that the functions of the larynx are not limited to protecting the lower respiratory tract but also to serve in the function of speech—an ability unique to human beings. Therefore, those who have undergone a laryngectomy have lost an important life-sustaining organ that deals with respiration and the protection of the lower

© Springer International Publishing AG 2017

T. Ifukube, *Sound-Based Assistive Technology*,

DOI 10.1007/978-3-319-47997-2_6

Fig. 6.1 *Left* Trachea-stoma;
right human speech organ

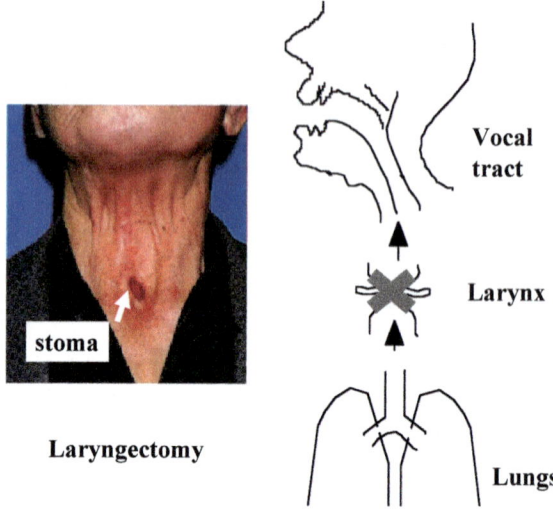

Laryngectomy

respiratory tract. At the same time, as such individuals have also lost the vocal folds inside the larynx, it becomes difficult for them to produce vocal sounds.

It is estimated that there are 20,000 such people in Japan, and approximately 1,000,000 worldwide. According to a survey conducted by members of the Japanese Laryngectomy Society, 86% of them are men and 14% are women, ranging in age from 32 to 98, with a mean age of 67 [1]. As a large percentage of the people in this survey are elderly, it would appear inevitable that the number of such elderly patients is gradually increasing in accordance with the rapid aging of developed countries as well as Japanese society.

Those who have had their larynx removed have completely lost the protective function of the respiratory tract. Specifically, as one essential role of the larynx is to stop food from entering the windpipe so that people will not choke, removal of the larynx makes it impossible to separate the esophagus from the trachea. Therefore, today's surgical technique is to completely close up the hole between the respiratory tract and the esophagus by stitching it up with the skin taken from the edge of the upper breastbone in front of the neck.

In this case, people who have had their larynx removed must breathe through a hole made in the throat called a "trachea-stoma" or simply a "stoma" (left figure of Fig. 6.1). This clearly shows that a laryngectomy not only removes the functions of the larynx, but also results in the need for a surgical operation to divide the lungs and the vocal tract. The unfortunate result of such an operation is that an individual is no longer able to exhale air into the vocal organs to resonate and produce sounds. However, voiceless consonant sounds, such as fricatives and plosives, which make use of the vocal tract as the sound source, can still be vocalized by changing the shape of the vocal tract and increasing the pressure inside the oral cavity. However, the vowels and voiced sounds that are produced from inside the larynx can no longer be vocalized after such surgery has been performed.

Nonetheless, even after such surgery, as some other voice-producing organs such as the tongue usually remain intact, it is still possible to produce the changes in resonation that are necessary for the production of vocal sounds. Therefore, those without a larynx can produce artificial vocal sounds by sending substitute sounds to the vocal tract as a replacement for the primary laryngeal tones that are normally produced in healthy vocal folds.

6.1.1 Various Speech Production Substitutes [2]

The first laryngectomy was actually performed in 1872. In the following year, researchers began to find a way to enable such individuals to regain the ability to produce vocal sounds without the use of a larynx. During the course of well over a century, numerous attempts have been made with the same goal in mind. However, individuals whose larynx has been removed have to vocalize in different ways. Until now, four vocalization methods, known as speech production substitutes, have been developed and are presently in practical use.

6.1.1.1 Esophageal Speech

The first of those methods is referred to as "esophageal speech" (Fig. 6.2a). When using this method, a person first attempts to accumulate around 15 ml of air in the esophagus. Then, by use of a belching-like technique to exhale the air, a voice-like sound is produced. Specifically, as the belching sound emerges from the mouth, a person is able to produce vocal sounds through changes in the shape of the vocal tract.

One drawback regarding this method is that it requires from six months to one year for people to adequately train themselves in its use. Moreover, there is great individual variation regarding the eventual degree of conversational ability that

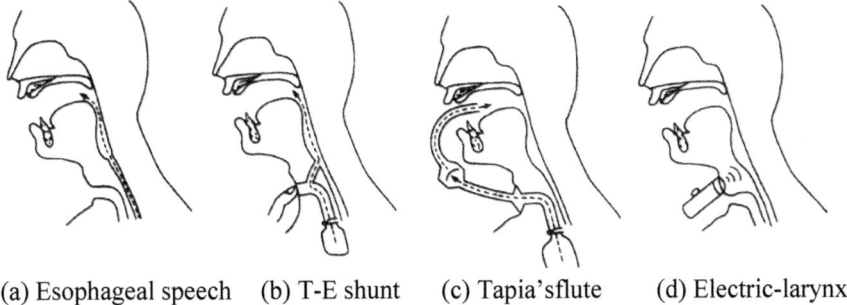

(a) Esophageal speech   (b) T-E shunt   (c) Tapia'sflute   (d) Electric-larynx

Fig. 6.2 Four conventional substitutes for speech disorders

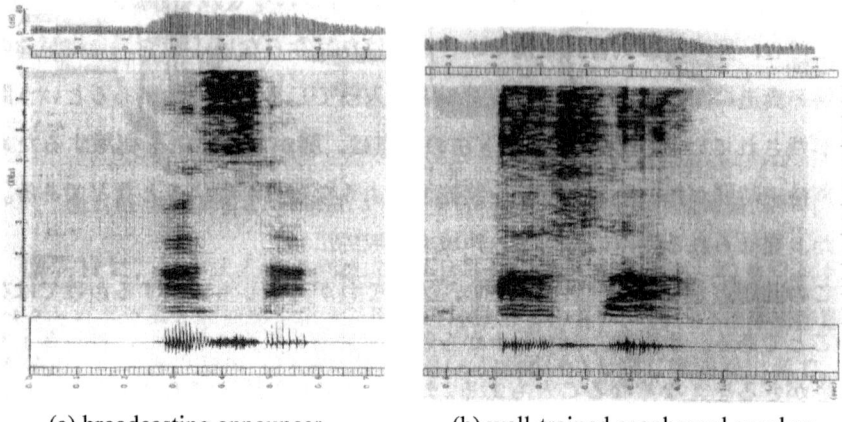

(a) broadcasting announcer   (b) well-trained esophageal speaker

Fig. 6.3 Comparison of spectrograms of the vocal sound /asa/ produced by **a** a broadcasting announcer with **b** those of a well-trained esophageal speaker

each person acquires. Lastly, as considerable training is also required to learn the correct way of swallowing air into the esophagus, people with considerable experience in this procedure are able to produce vocal sounds of superior clarity than new learners.

Figure 6.3 shows spectrograms for the esophageal speech sounds produced by an announcer and a well-trained esophageal speaker. It is clear that when using esophageal speech sounds it is difficult to find the correct pitch. Further compounding this difficulty is the fact that strong noise-like elements appear in the high-frequency ranges.

Laboratories that focus on medical-assistive devices are developing and researching instruments aimed at improving sound quality by clarifying as well as amplifying sounds. However, as the potential for speech production depends on the formation of the lower pharynx following a laryngectomy, it is almost impossible for many people to learn how to acquire a new voice despite considerable practice. More specifically, even if a person succeeds in learning the technique, their voice quality remains poor. Even for experienced individuals, the periods of time during which they can produce speech are short because there is a limit to the quantity of air that can be taken into the esophagus at one time. In fact, roughly three years of practice are required to adequately learn this technique, which leads to the unfortunate result that 40% of those who attempt to learn it are ultimately unsuccessful.

One problem with general speech production substitutes is the fact that it is difficult to distinguish voiced from voiceless consonants. However, those who have acquired an excellent esophageal speech ability can distinguish voiced and voiceless consonants with considerable accuracy. One drawback to this method, however, is that the necessary training, which is both demanding and exhausting, makes it particularly difficult if not impossible for the elderly to master.

6.1.1.2 Trachea-Esophageal Shunt

The second speech production substitute is known as the "trachea-esophageal speech production method," or the T-E shunt method for short (Fig. 6.2b). In this method, a shunt or bypass is surgically created, forming a hole in the trachea (T) and the esophagus (E), known as the stoma. When the stoma is closed using a finger, the air moves from the hole to the esophagus. An alternative glottis, known as a pseudo-glottis, then vibrates to produce sound, thus allowing the person to speak by use of this T-E shunt method.

However, one difficulty in using the stoma is that a person must always first close the hole with her or his finger to be able to speak. Additionally, there is a danger that food might accidentally enter the trachea because the same holes lead to both the trachea and the esophagus. On the other hand, one advantage of the T-E speech production method is that its use is not limited or constrained in any way by the need for special equipment or materials; in short, users can continue speaking this way over a long period of time simply by making use of the expiration of air from their lungs. Another advantage is that such a method allows for the production of a wide range of vocal sounds created at various intensities. In addition, it also offers a great advantage in that the smell function is reserved.

Unfortunately, patients undergoing such surgery cannot be guaranteed that it will always be successful. For example, in some cases it may become necessary to eventually close the hole if too much saliva or food leaks into the trachea through the shunt. Other possible issues are that excessive friction resulting from constant use of the shunt may cause high pressure in the trachea. If this should occur, two unfortunate effects would likely occur: more difficulty in speaking, combined with inferior sound quality due to irregular vibrations emitted from the sound source. Lastly, the very fact that use of this method necessarily occupies one of the user's hands can, at times, prove troublesome.

On a more positive note, since the success of Singer and Blom in designing an instrument with a check valve that prevents saliva and food from leaking into the trachea [3], many other instruments and methods have been designed with the same goal in mind. Furthermore, a device fitted with a reed that can be inserted into the trachea-esophageal passage has been developed with the dual function of securing the sound source in addition to preventing the accidental swallowing of food into the trachea. Reports regarding the efficacy of this device claim that pronunciation is significantly improved.

6.1.1.3 Whistle-Type Artificial Larynx (Tapia's Flute)

The third speech production method is called the reed-based artificial larynx (Fig. 6.2c). It is also known as "Tapia's flute," which now includes a number of improved versions. With this type of artificial larynx, a rubber membrane or metal board is used to cover the opening of a device that is attached to the stoma in order to extract the air that is exhaled. As the membrane or board vibrates during exhalation, the vibrating sounds then enter the mouth through the pipe. In this way, the sound source is delivered to the vocal tract. If the vibrating place is like a rubber membrane, through practice, a user can learn to control the pitch frequency by adjusting the membrane's tension and shape. However, despite reports that some exceptional users are able to adjust their voices enough to span the range of one full octave, the voices generally sound monotonous to the listener.

6.1.1.4 Electrical Artificial Larynx

The fourth and final practical vocalization method is the electrical artificial larynx, simply called the "electro-larynx" (Fig. 6.2d). The term electro-larynx is generally applied to artificial larynxes that use an electromagnetic vibrator as the sound source. To enable a user to speak, this method applies a vibrator to the lower jaw, which sends a vibrating sound into the vocal tract through the soft subcutaneous tissue.

One advantage of this system is that it can be easily learned. Additionally, as it does not require the user to exhale, people are free to speak for as long as they like without any concern regarding respiration. However, some of its shortcomings are that the voice quality is rather poor and tends to have a buzzer-like quality to it. Moreover, as some vibration sounds leak out from the vocal tract, they may cause interference for the listener. As with the whistle-type artificial larynx, it can also be inconvenient for the user to always have one hand occupied with the device. A further drawback is that this system cannot be used to send vibrations to the vocal tract for those who have serious scar tissue due to radiation therapy. As regards the sound source, there are reports of users being able to control both accent and intonation by applying a finger to the device. However, it may be difficult to use this approach during a conversation without appearing unnatural.

Nevertheless, this method has recently become rather popular due to its ease of learning. In the United States, people who can communicate through esophageal speech are even instructed to carry the electro-larynx around with them. Considering this situation, there is great expectation to improve the effectiveness of this assistive device by adding naturalness so it sounds like a human voice.

6.2 Vocalization Mechanism of Talking Birds and Its Application

The ability to produce vocal sounds is not limited to human beings. Among the animals that can produce vocal sounds are certain birds, including mynahs, parrots, and parakeets. These birds are especially loved as pets for their uncanny ability to mimic human speech. In particular, a mynah bird's speech sounds quite smooth and natural. For researchers of sound-assistive technology, it is a mystery as to why these birds, whose mouths and ears are shaped in a totally different way from those of humans, are capable of distinguishing and vocalizing certain human words.

Therefore, it is possible that an examination of the vocalization mechanism of mynah birds may lead to new ideas for the development of the artificial larynx. Moreover, it could offer a new perspective on how to improve the synthesis of speech sounds.

6.2.1   Vocalization Mechanism of Mynah Birds

As a concrete example of the mynah bird's ability, it is worth mentioning the author's experience with a mynah bird. After learning how to pronounce a number of words, the bird was able to imitate the voices of various people. However, more noteworthy than such simple word repetition, was its ability to utter such seemingly appropriate responses as, "Ouch! That hurts!" after falling off its perch. Furthermore, its word utterances were always accompanied by a conditioned reflex, such as chirping, twittering or even imitating the ringing of a telephone. Since the human auditory sense enables us to recognize speech sounds whose waves differ from our own, such as in the case of mynah birds, conducting research on the speech sounds of mynah birds may provide a clue towards solving the difficult problem of recognizing the words of unspecified speakers. It is hoped that the knowledge gained by such research on mynah birds will provide valuable information leading to the development of artificial larynxes. In this section, the author would like to explain a little more precisely the speech-production mechanism of the mynah bird and also look at why the mynah bird's voices might sound so natural to a human.

Figure 6.4 is a model of a mynah bird's speech organs. Although a mynah bird has nostrils, the shape of its nose differs greatly from that of humans. Moreover, they have a beak instead of lips and their tongues are narrow and thin. At the dividing point between the trachea and the bronchial tubes, there are two "syrinxes" that correspond to vocal folds. Hirahara and the author conducted an experiment using vocal sounds produced in helium-oxygen gas to examine the differences in the vocalization mechanism between human beings and mynah birds [4].

In this experiment, nitrogen, which comprises about 80% of air, was replaced by helium. The helium-oxygen gas was then fed into the cage and both a human being and a mynah bird vocalized sounds (Fig. 6.5). As helium is lighter than nitrogen, the speed of the sound increased. Given that sound speed equals frequency times wavelength, and given that the frequency in helium-oxygen gas remains constant, the wavelength changes. This means that the resonance frequency changes even within the same acoustic tube. More specifically, the longer the wavelength, the more the resonance frequency shifts to the higher end of the frequency range. In the case of the human voice, the changes or shifts occur in the formant frequencies.

When we analyzed the spoken vocal sounds (top-left figure of Fig. 6.6) of humans (/i/) and observed the spectrum, the second and third formants had clearly

Fig. 6.4 Speech organ of the mynah bird

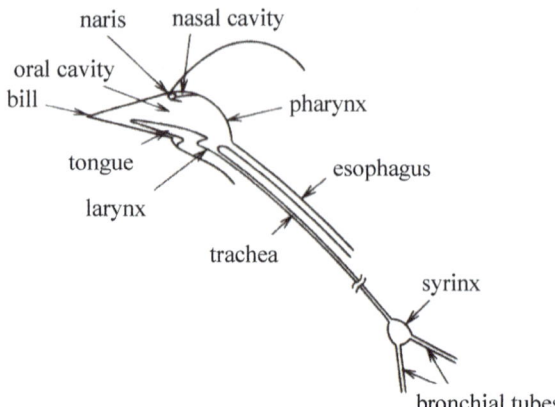

Fig. 6.5 A mynah bird and its tutor inside a cage filled with 80%He (helium) - 20%O$_2$ (oxygen)

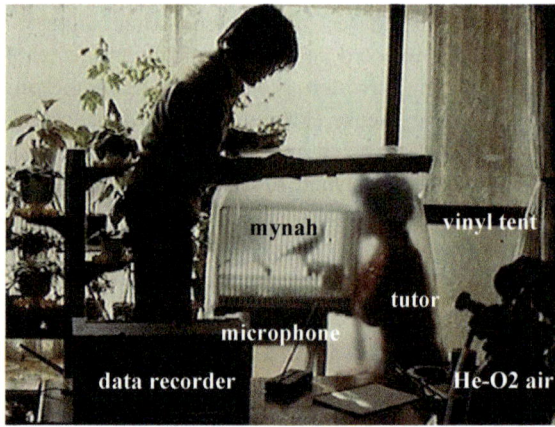

shifted to the high-frequency band. As a result, the sounds resembled those of "Donald Duck" from the cartoon. However, in the case of the mynah bird, even although the third formant shifted, the frequency of the second formant did not change at all, as seen in bottom-right of the figure. In addition, as the spectrum shows, the first formant was extremely small whereas the second formant was large.

This leads to the assumption that the second formant is not produced by resonance. Consequently, it is not appropriate to use the word "formant;" rather, it ought to be called the "peak" of the spectrum envelop. It goes without saying that the second formant is the most important element in characterizing vowel sounds. This peak that corresponds to the second formant does not change in the case of helium-oxygen gas, but does change according to vowel sounds. More concretely, when the sound /i/ is released, the peak moves close to 2000 Hz; when the sound /a/ is released, it moves to around 1500 Hz. Obviously, it can be assumed that the mynah bird itself is controlling the frequency that corresponds to the second formant, as schematically represented in Fig. 6.7.

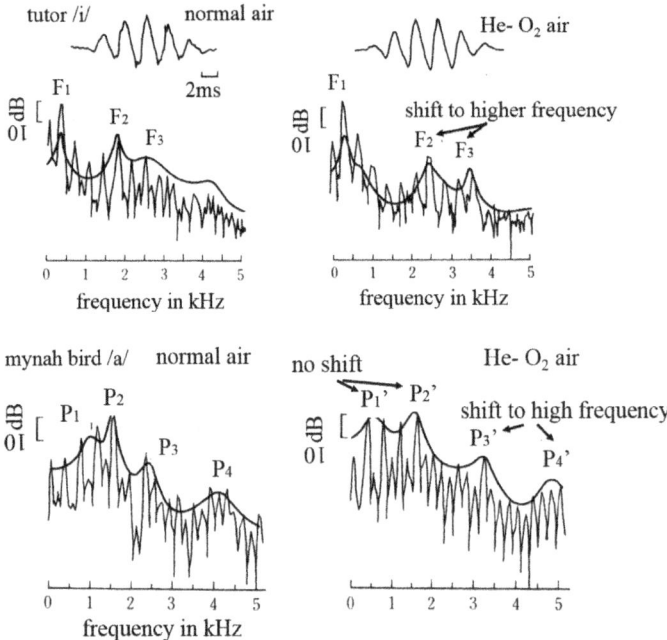

Fig. 6.6 Four spectral patterns, *left* in normal air, *right* He-O$_2$ air, *top* tutor's vowel /i/, *lower* Mynah bird's vowel /a/

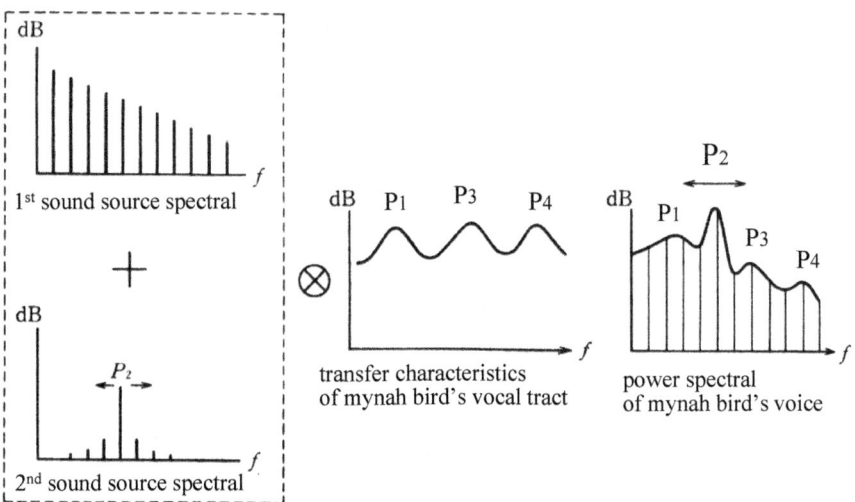

Fig. 6.7 Schematic representation of the speech-production mechanism of the mynah bird

Calculating the cross section of the vocal tract of a mynah and examining its syrinx or vocal organ tells us the following: First, as the left figure indicates, the two syrinxes seem to be controlled separately. One syrinx produces sounds at about 200 Hz—a level similar to that for human vocal folds; the other syrinx seems to release the elements that directly correspond to the second formant. When a mynah bird produces sounds, these two sound sources are combined, with sounds at around 200 Hz changing as a result of being influenced by the resonance characteristics of the first and third formants—a process that is unique to mynah birds. This spectrum can be seen on the right side of the figure.

Figure 6.8 shows the area around the syrinxes of a mynah bird. There is one tympani form membrane (outer lip in the figure) on both sides of the trachea that is separated from the beak (upper direction of tracheal lumen in the figure). Within each membrane, there are three bones indicated in the figure by "A_1, A_2, A_3" and a sac covered with a thin membrane (internal lip in the figure) in the center. A mynah bird cannot utter a sound if a hole is made in the sac because this would cause the sac to deflate. Both sides of the tympani form membrane are pushed up by the movement of the three bones and strike the sac, which is filled with air like a tympani. In addition, although the striking frequency is different on the right and left sides, the spectrum is finally converted to that of the right side due to the effect of the vocal tract along the way.

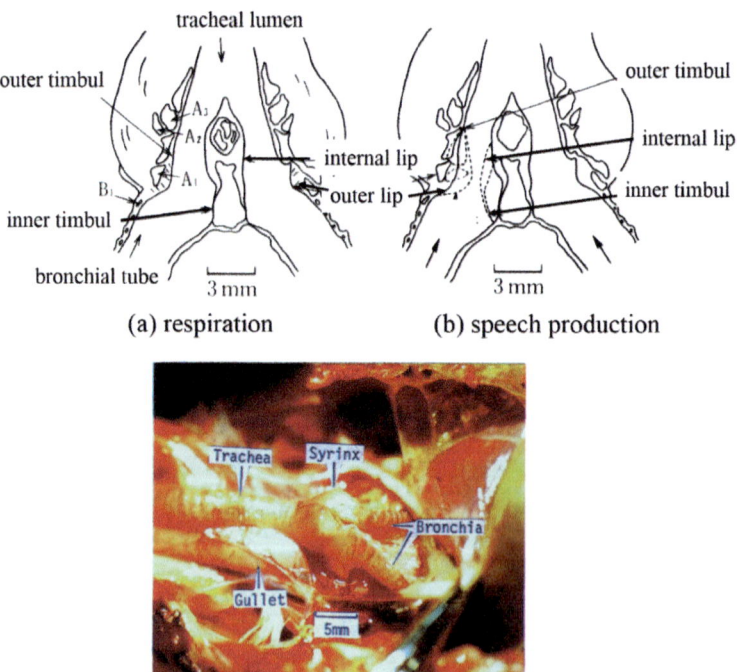

Fig. 6.8 Two syrinxes of the mynah bird. *Upper* anatomy chart of the syrinxes, *lower* photograph of syrinxes

6.2.2 Naturalness of the Mynah Bird's Imitation Voice

In order to create a natural, human-sounding voice, two elements are necessary, the first of which is fluctuation. More specifically, it is the subtle fluctuations of waves over time combined with a fluctuating pitch frequency that helps to create humanlike sounds. The second necessary element for producing a human-sounding voice is metrics, including such factors as accents and intonation. In fact, the role of intonation is particularly essential. The reason why so-called mimicking birds possess such a rare ability is because they are able to faithfully imitate this fluctuation and intonation.

The top graph in Fig. 6.9 shows the pitch frequency pattern of a mynah bird uttering the Japanese phrase /kawaiine/, which in English means "You are cute." It is clear that a mynah bird's pitch and intonation are surprisingly similar to those of a human being (bottom graph), thus producing what to us sounds like a human voice.

Although it is relatively easy to learn to use an electro-larynx, it nonetheless requires that the person accept the need to use an artificial device for communication. A further drawback is that the electro-larynxes in use today are relatively poor at voice adjustments, thus they produce sounds characterized by a rather monotonous pitch mixed with other forms of miscellaneous noise during conversation. As previously described, we have learned that mynah birds produce humanlike vocal sounds by imitating fluctuation and intonation; this knowledge may be used as a clue in present-day research aimed at improving the quality of today's artificial larynxes by producing sounds that more closely resemble those of a normal human voice.

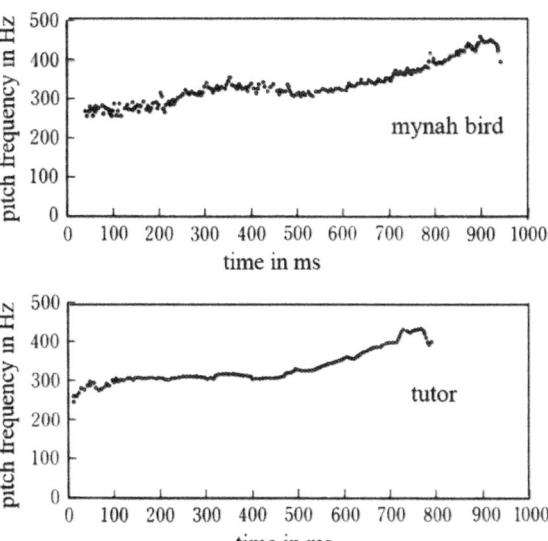

Fig. 6.9 Pitch patterns of /kawaiine/ ("You are cute") extracted from a tutor and a mynah bird

In the next section, the author will explain how this knowledge gathered on the mynah bird helped to stimulate research on the development of a new kind of electro-larynx that contains the essential elements of intonation and fluctuation.

6.3   Electro-larynx with Intonation and Fluctuation

6.3.1   Pitch-controlled Electro-larynx

In the creation of vocal sounds, an important role is played by the flow of air from the lungs to the larynx. More concretely, by controlling the way air is exhaled, the lungs and larynx determine the volume, accent and rhythm of a sound. In view of this, it is conceivable that sounds produced by an artificial larynx could likewise be determined by controlling how the lungs exhale air.

For example, it is possible to produce vocal sounds by attaching a sensor to the trachea-stoma to convert information regarding the characteristics of the exhalation into a vibratory frequency that is then applied to the throat. As exhaling strongly causes the vibrator to vibrate at a high frequency and exhaling lightly results in low-frequency vibration, such an approach would facilitate the control of intonation. In this way, it is assumed that a pitch-controlled electro-larynx can be realized, as shown in Fig. 6.10.

However, one troublesome problem regarding intonation is that it tends to change too much when there is merely a slight change in expiration pressure, but fails to change sufficiently even when the expiration pressure undergoes considerable change. However, there is an ideal equation that can be used to convert expiration pressure to pitch frequency and this function was applied in an effort to ascertain the most suitable parameters for several laryngectomees. Uemi and the author conducted an experiment by taking two factors into account: the subjective evaluation of each test subject of how easy it was to control the exhalation pressure from an artificial larynx; and the degree of naturalness associated with the vocal sounds it produced [5].

Fig. 6.10 Block-diagram of
p a pitch-controlled
electro-larynx

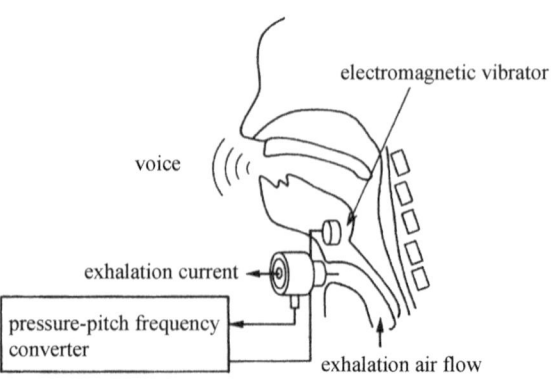

The electro-larynx used in the experiment was set up in such a way as to prevent vibrations from beginning prior to the exhalation pressure having reached a certain value. Specifically, this was done to control the timing, so that a user would be able to turn the device on and off in accordance with the beginning and completion of each period of vocal sound generation. Given that the physiological voice range of normal adult males is from 60 to 500 Hz, the equation to convert expiration (P [cmH$_2$O]) to sound pitch (f [Hz]) was fixed as f = k(P − P$_0$) + f$_0$, P$_0$ = 1 cmH$_2$O, f$_0$ = 60 Hz, so that sounds under 60 Hz would not be generated. By referring to the value for those with normal hearing, the slope k was fixed at five parameters. As the air current's resistance is nonlinear, an exponential conversion function was used to convert the air current into expiration pressure.

Five subjects who had undergone laryngectomies were asked to vocalize the words /aoi umi/ (meaning, "blue ocean") by controlling the exhalation pressure five times for each parameter. At the same time, they were asked to evaluate the ease of control on a scale of 1–5. The naturalness of the vocalized speech sound was based on pair comparisons.

The top figure of Fig. 6.11 shows the average score for five hearing subjects. The vertical axis represents the degree of naturalness and the standard deviation, while the slope of the equation is indicated by the horizontal axis. Based on this diagram, the degree of naturalness and ease of control received the best scores when the slope (k) was 25/cmH$_2$O. Moreover, using this slope, the feeling of resistance felt by the subjects corresponded to when they were breathing through overlapped masks.

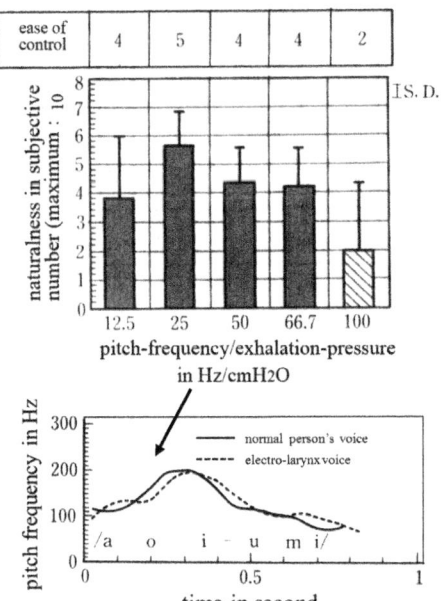

Fig. 6.11 *Top* Ease of pitch control and relation between naturalness and pitch-frequency/exhalation-pressure, *bottom* comparison of pitch frequency (/a o i u mi/"blue sky") between a normal person's voice and electro-larynx voice

Fig. 6.12 Photograph of a pitch-controlled electro-larynx

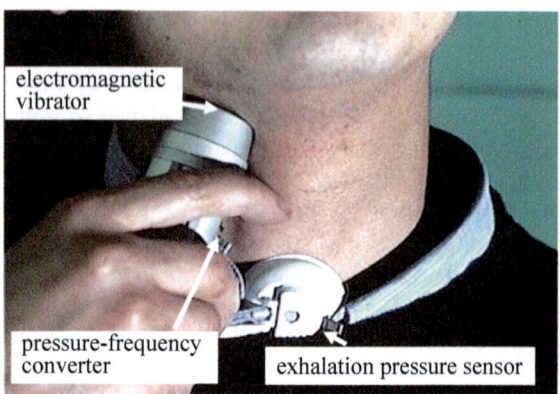

electromagnetic vibrator

pressure-frequency converter

exhalation pressure sensor

The bottom part of Fig. 6.11 indicates the resultant pitch pattern when a person without a larynx vocalized /aoi umi/ using the optimal parameters. It is clear that the pitch pattern closely approximates that of normal subject, as shown by the solid line.

After completing this basic research, a design was created for an electro-larynx consisting of an expiration sensor, a main body and a vibrator. Based on this design, a device was developed for practical use with the cooperation of an electronics manufacturer (DENSEI LTD.) and Hokkaido Industrial Research Institute in 1998. Figure 6.12 shows a photograph of a pitch-controlled electro-larynx with sales of more than 10,000 units in Japan. However, the electro-larynx sound is still not a human-sounding voice. It was assumed that fluctuations in speech waves need to be added to the electro-larynx voice. Furthermore, as this method requires the use of a hand, it is still somewhat restricting. Many users have asked us to design a hands-free electro-larynx so that they can use it in daily life and at their office without needing their hands to control the electro-larynx. As described in the next two sections, we have tried to solve these two problems by designing a hands-free electro-larynx that can produce pitch fluctuations.

6.3.2 Electro-larynx with Vocal Sound Fluctuations

The author and his colleagues conducted preliminary research regarding the relation between voice naturalness and the sidebands of the voice spectrum [6]. Their results offer clear evidence that approximately 32 pitch waves and a lateral spectrum of ± 18 Hz around the harmonics of the voice are needed to maintain a natural voice quality. As can be seen in the left part of Fig. 6.13, in the electro-larynx used in the experiment, a small speaker was inserted into a cylindrical pipe from which sounds were emitted into the oral cavity through a hole in the pipe.

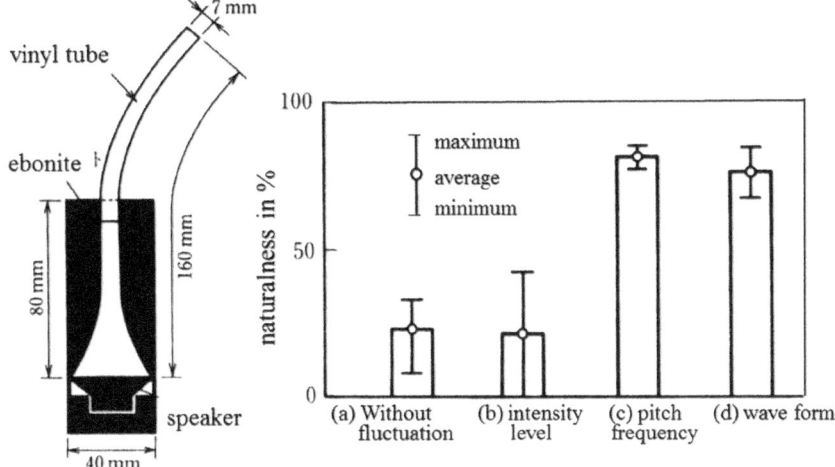

outline of the artificial larynx

Fig. 6.13 Naturalness comparison for four voices produced from a tube-type artificial larynx

Next, as shown in the right part of the figure, we examined how the degree of naturalness improves by adding fluctuations. In this experiment, four vocal sounds were used as the sound source waves: a pulse train (a) and three processed sound sources from an original sound source (b–d). The original source was made from the residual signals calculated by an LPC analysis of a normal sustained vowel /a/ with 32 pitches.

(a) Sound source created by inputting a pulse train into a LPF (−12 dB/oct with a cut-off frequency of 90 Hz);
(b) Sound source created by residual signals with intensity fluctuation;
(c) Sound source made from residual signals with partial wave fluctuation;
(d) Sound source produced from residual signals with full wave fluctuations;

The right part of the figure shows the results of the experiment. The horizontal axis represents four kinds of sound source, while the vertical axis represents naturalness. Judging from this diagram, it is clear that sounds (c) and (d) were characterized by a high degree of naturalness, whereas the quality of naturalness for sounds (a) and (b) was low.

This finding shows that fluctuation of sound sources is indeed necessary to maintain vocal sound quality. Furthermore, the results indicate that there is no significant difference in the naturalness of sounds (c) and (d). Therefore, it can be said that the naturalness of electro-larynx vocal sounds is sufficiently improved by the use of sound source waves having wave fluctuations that are composed of amplitude and pitch fluctuations.

On the other hand, in wave fluctuations, two kinds of fluctuations are always observed in the steady parts of normal sustained vowels. One is amplitude

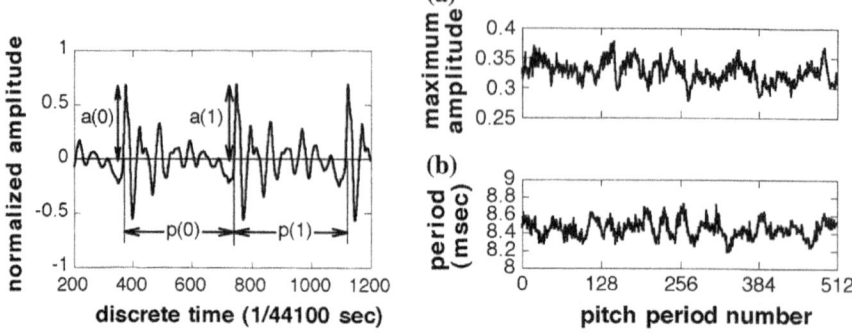

Fig. 6.14 *Left* Extraction of amplitude sequences (AS) [a(n)] and period sequences (PS) [p(n)], *right* examples of **a** AS and **b** PS obtained from one of the speech samples

fluctuation, defined as cyclic changes in the maximum peak amplitudes. The other is period (pitch) fluctuation, defined as cyclic changes in the pitch periods. Aoki and the author investigated which fluctuation—amplitude or pitch fluctuation—is dominant in determining the naturalness of the vowels based on quantitative descriptions of amplitude and period sequences obtained from normal sustained vowels and psychoacoustic experiments.

These fluctuation sequences composed of maximum peak amplitudes or pitch periods were extracted successively from 512 consecutive pitch periods in the steady part. As shown in the left side Fig. 6.14, the amplitudes of the speech sample were normalized to vary from −1 to 1, which corresponds to −32,768 to 32,767 in 16-bit quantization. From 512 consecutive pitch periods, the amplitude sequences (AS) denoted by a[n] (n = 0, 1, ..., 511) and the period sequences (PS) denoted by P[n] (n = 0, 1, ..., 511) were successively extracted. The examples of the extracted AS and PS are shown in figures (a) and (b) of the right side. The frequency characteristics of the AS and PS waves are shown in Fig. 6.15a, b, respectively.

The results of the frequency analysis indicated that the frequency characteristics seemed to be subject to a spectral 1/f power law. The value of β in $1/f^\beta$ is 0.99 for the AS and β = 0.96 for the PS. To investigate the possibility that the frequency

Fig. 6.15 Mean frequency characteristics of **a** AS and **b** PS obtained from all speech samples. Here, the value of β is close to 1; β = 0.99 for the AS and β = 0.96 for the PS

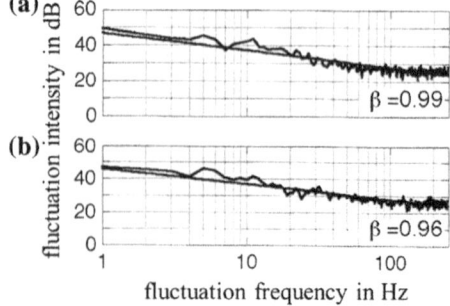

Fig. 6.16 Results of psychoacoustic experiments: **a** original is continuous male speech, and **b** original is continuous female speech

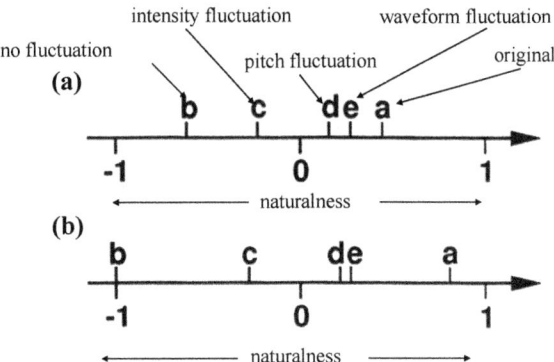

characteristics of the fluctuation sequences influence the voice quality of sustained vowels, Aoki and the author conducted psychoacoustic experiments. The degree of naturalness was evaluated by using the pair-comparison method for the four sequences with different β. The four sequences were spectral $1/f^0$ (white noise), $1/f$, $1/f^2$, and $1/f^3$ sequences. The experimental results indicated that the subjective voice quality of synthesized sustained vowels is highest for spectral $1/f$ values among the four sequences, and that the voice quality of synthesized vowels with PS is higher than that with AS.

Furthermore, we investigated the naturalness, based on the psychological distance method, of (a) original vowels, (b) vowels without any fluctuation, (c) vowels with AS, (d) vowels with PS, and (e) original vowels with wave fluctuation. The results indicated that the voice quality of the vowels with PS is very close to that with wave fluctuation. Especially, in male speech, the voice quality of the vowels with PS is located near to that of the original vowels, as shown in Fig. 6.16.

Based on these findings, we have designed a new mode of the electrical artificial larynx that can produce sound sources with PS (i.e., pitch fluctuation). The electro-larynx with pitch fluctuation has been manufactured as "Your Tone II" since 2013 in Japan.

6.3.3 Wearable Hands-Free Electro-larynx

Several attempts have been made to satisfy demands from users for a hands-free electro-larynx, some of which are shown in Fig. 6.17 [7–10]. However, it is still difficult to manufacture them, mainly due to the complexity of doing so, and they remain difficult to use. However, Hashiba and his colleague designed a hands-free electro-larynx that can be attached to the user's neck. They identified four problems that remained to be solved. One is how to attach an electro-larynx around a user's neck. The second is how to make the vibrator small and light. The third is how to pick up the electro-larynx voice and amplify it. The fourth is how to improve an

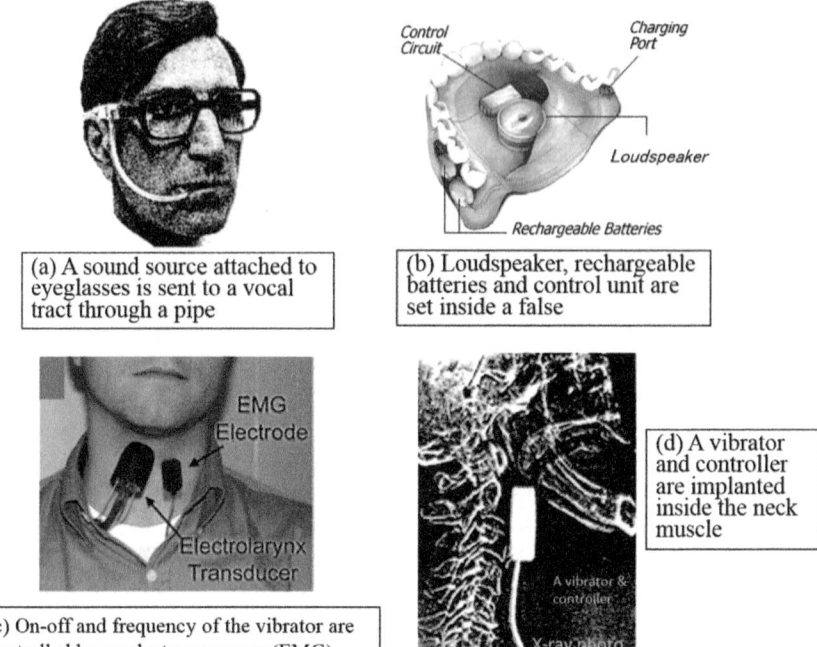

(a) A sound source attached to eyeglasses is sent to a vocal tract through a pipe

(b) Loudspeaker, rechargeable batteries and control unit are set inside a false

(c) On-off and frequency of the vibrator are controlled by an electromyogram (EMG)

(d) A vibrator and controller are implanted inside the neck muscle

Fig. 6.17 Some attempts at a hands-free electro-larynx

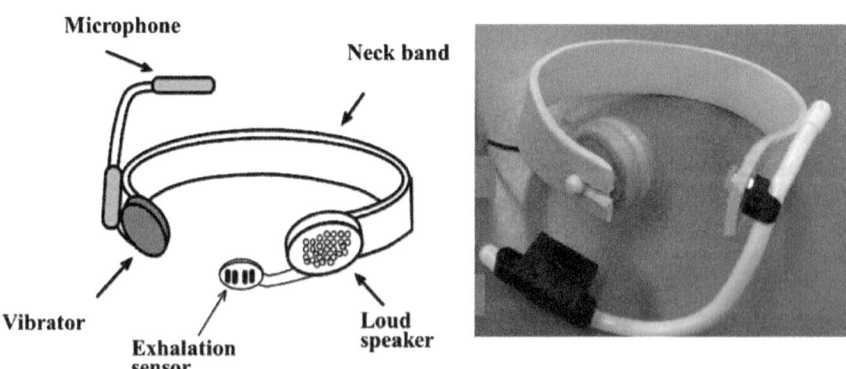

Fig. 6.18 *Left* Conceptual design of a hands-free pitch-controlled electro-larynx, *right* a neckband, a vibrator and an exhalation sensor for a hands-free pitch-controlled electro-larynx

exhalation sensor in order to adjust the hands-free electro-larynx. The left part of Fig. 6.18 shows a conceptual model of an electro-larynx that would solve these problems, and the right shows a neckband, a vibrator and an exhalation sensor for the electro-larynx.

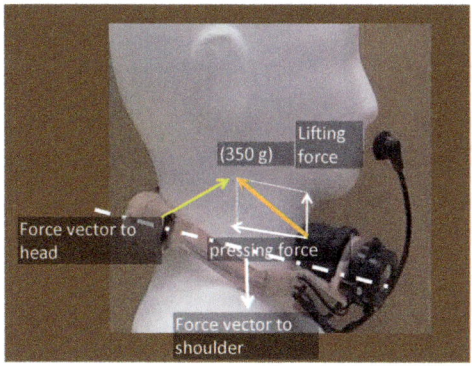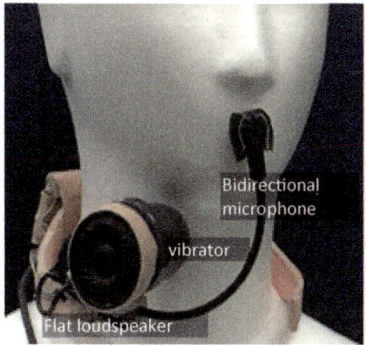

Fig. 6.19 Design of a hands-free electro-larynx

For the neckband, after comparing various materials, a thermoplastic brace was selected. It becomes soft at a temperature of about 70 °C, while being hard at body temperature, so it is suitable for adjusting to individual differences in shape and hardness of the neck tissue. Furthermore, the shape and the hardness of the thermoplastic brace were determined so the vibrating sound could enter the vocal tract even if the head and neck of the user moved while talking. We designed the neckband by considering force vectors, as shown in the left part of Fig. 6.19 [11].

They also made a smaller vibrator (32 mm diameter × 19 mm thick and 32 g) and a much more sensitive flow sensor than the conventional one. The vibrator and the sensor can be on the thermoplastic brace. The top right figure shows a prototype of the hands-free electro-larynx being worn. A microphone was designed to pick up only the voice produced from the lips by using bilateral directional characteristics, as shown in the right figure.

From usability tests, it was confirmed that electro-larynx voices can be heard more naturally than a conventional device, and that the activity of the users significantly improved in daily life. Furthermore, by adding a 1/f pitch fluctuation to the electro-larynx, the voice quality exceeded that of, at least, a mynah bird's imitation.

6.3.4   Methodology of Artificial Larynx Implants

The ideal artificial larynx implant is one that would allow users to speak freely over a long period of time whenever they desired to do so. In order to determine the potential offered by such an artificial larynx implant, Min and her colleagues have examined electromyograms (EMG) of the vocalizing muscle that can remain following a laryngectomy, specifically, to what extent can this muscle still be used to turn the sound source on and off, to control whether the sounds are voiced or voiceless, and to control pitch frequency ([12]).

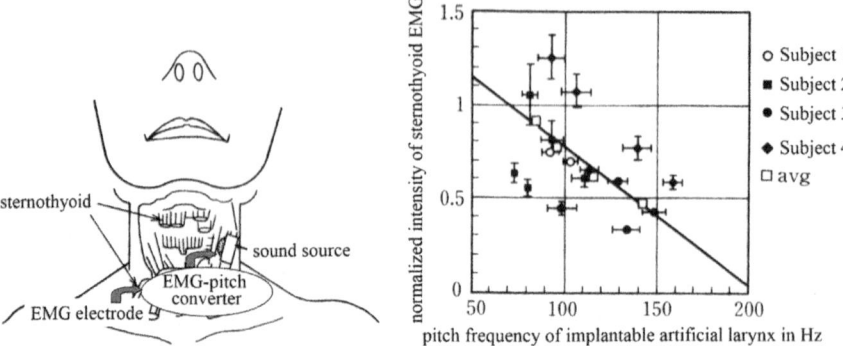

Fig. 6.20 *Left* Design concept for an implantable artificial larynx, *right* EMG level versus pitch frequency of an implantable artificial larynx

One of larynx muscles that can be left following a laryngectomy is the sternohyoideus that connects breast bone and hyoid bone, as shown in the left side of Fig. 6.20. Originating from the breast bone and inserted into the surface of the hyoid bone, the action of the sternohyoideus is to depress the hyoid bone. It is also said to be used in vocalizing voice pitches that are lower than usual. By preserving the sternohyoideus, the muscle can be used to control an electronic artificial larynx.

In a preliminary study, by using electrodes attached to the muscles around the sternohyoideus of subjects (HK, YI, and HN in the figure) with normal speaking ability, an examination was conducted to ascertain their ability to use the muscle to turn the sound source on and off, to control whether the sounds are voiced or voiceless, and to control the pitch frequency. In this experiment, after the subjects heard the music scale sounds *do* (C_3), *re* (D_3), *mi* (E_3) and *fa* (F_3) on an electric piano, they were asked to consistently use the sound /i/ to reproduce the same four scale sounds. The right figure shows the pitch frequency, when it was followed by the sound pitch, and the scale of the electrodes on a normalized sternohyoideus muscle at that point in time. Although there was considerable individual variation in the sounds produced by the experimental electro-larynx, the average possible change in the pitch frequency was from 85 to 143 Hz.

An 80-year-old man who had undergone a laryngectomy, but whose sternohyoideus had been severed, participated in the next step of this experiment. Eight months after the removal of his larynx, the subject was asked to vocalize /a/ by using his sternohyoideus to control the experimental electro-larynx. Although the sound quality was unstable, it was confirmed that he was able to vocalize to some extent. In view of this finding, it would seem plausible in the future to develop an artificial larynx implant capable of intonation that is controlled by electrodes attached to the remaining portion of the larynx muscle.

6.4   Voice Synthesizer for Articulation Disorders and Dysphonia

To assist articulation disorders, the author conceived the idea of a speech synthe-sizer that can be controlled by a finger, as if a user were playing a musical instrument. One of the author's associates, Yabu, had some difficulties controlling his speech organ because he had been suffering from a kind of muscle disease, designed a synthesizer for himself. He had been disappointed that ordinary speech aids did not work in real time, or failed to produce nonverbal information such as intonation and emotional expressions.

Our method was modeled after the speech-production mechanism of a well-known ventriloquist, who can produce normal speech without moving his lips. By observing and measuring the movements of his lips and speech sounds, it was clear that the ventriloquist could produce bilabial consonants such as /pa/, /ba/, and /ma/ by rapidly moving the tongue without closing his lips. Based on this ven-triloquism vocalization mechanism, it was hypothesized that the ventriloquist quickly moves his tongue so that the /a/of /pa/ can be produced after placing the tongue against a front tooth; an articulation point of the consonant /t/, in the case of the ventriloquist's /pa/, as shown in Fig. 6.21. Humans can hear only /pa/ of the ventriloquist's /p/ even if the sound of the consonant /t/ exists. Almost the same mechanism as /pa/ was ascertained in the case of /ba/ and /ma/. This result means that /pa/, /ba/ and /ma/ could be made hearable solely by tongue movement inside the mouth; by changing formant frequencies [13].

The idea of our speech synthesizer is mainly based on the fact that humans recognize the bilabial consonants produced by a ventriloquist. In a new speech-production method for articulation disorders, first, the parameters of first

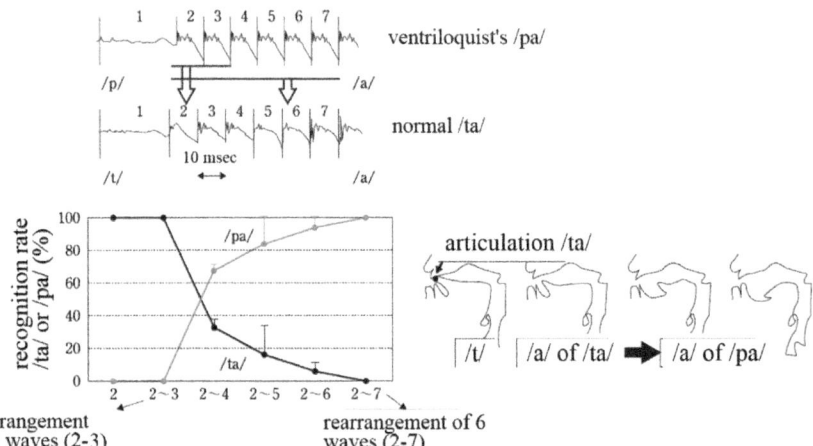

Fig. 6.21 A ventriloquist quickly moves the tongue so that /a/ of /pa/ can be produced after placing the tongue at an articulation point of /t/

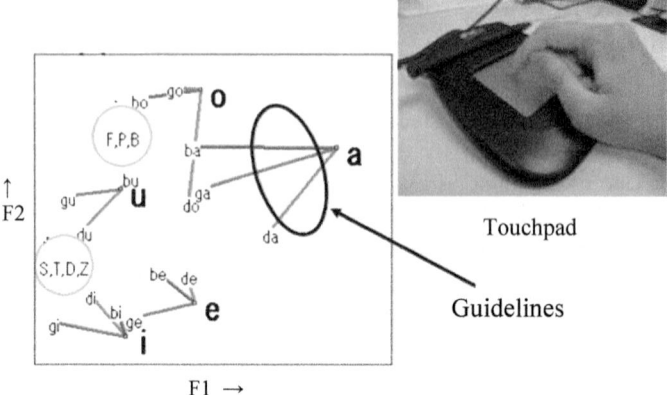

$F1 \rightarrow$

Fig. 6.22 A touchpad (*right*) where F1 and F2 frequencies were two-dimensionally assigned to the plane (*left*). Guidelines are drawn and noise components are arranged on the F1-F2 plane

formant frequency (F1) and the second formant frequency (F2) of a formant synthesis software could be controlled by a user's index finger placed on a touchpad. The F1 and F2 that correspond to the position of the tongue were two-dimensionally assigned to a plane of the touchpad (Fig. 6.22) [14].

A prototype model was designed on a personal computer using formant synthesis software to produce the optimal speech synthesis method for articulation disorders as shown in Fig. 6.23. In this method, four bandpass filters indicated in Fv1, Fv2, Fv3, Fv4 in the figure are used to create formants, although central frequencies of Fv3 and Fv4 were fixed. Therefore, the bandpass filters Fv1 and Fv2 correspond to the first formant frequency (F1) and the second formant frequency (F2), respectively. The arrangement of the five Japanese vowels on a plane was determined based on the relationship between the positions of the tongue and recognized vowels.

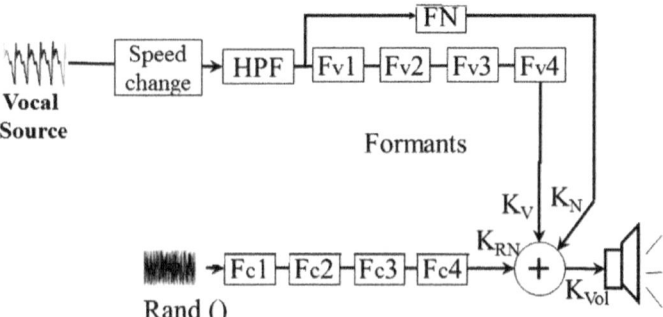

Fig. 6.23 Formant cascade model for a voice synthesizer controlled by a finger. Three noise components are added in order to produce /h/, /s/ and /ts/

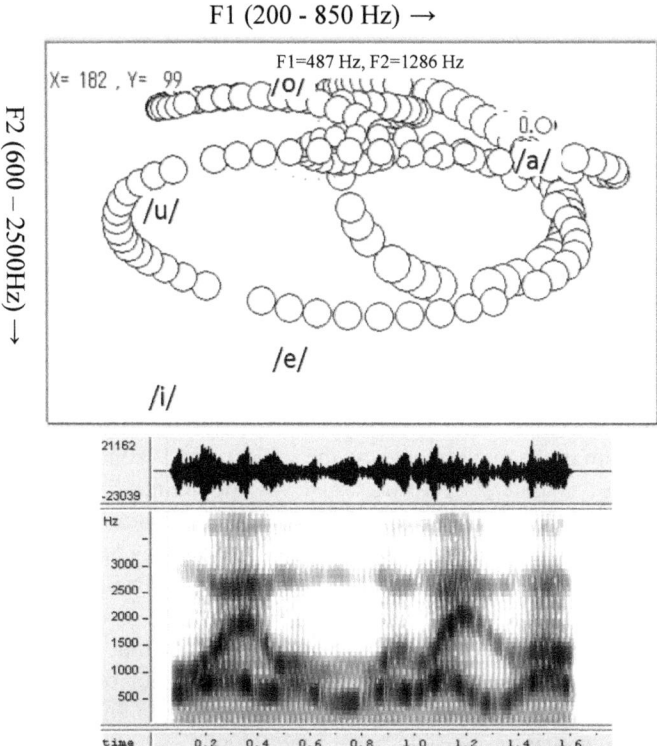

Fig. 6.24 An example (/ohayo-gozaimasu/, "good morning") of reproduced sentences. *Top* Trajectory of a finger on the touchpad, *bottom* time spectral pattern of the sentence

In order for users to easily find the formant transitions to produce the desired consonants only by changing the formant transitions, guidelines were superimposed on the pen-tablet as shown in the left of Fig. 6.22. The direction and length of each guideline were determined by referring to the formant transition locus of each consonant. The starting point and the formant transition locus were decided according to an "Expanded Locus Theory" which we modified based on "Locus Theory" [15].

In the evaluation test for sentence recognition, continuous speech sounds were synthesized by tracing the touchpad using the index finger, as shown in the top figure of Fig. 6.24. After a few hours training to produce some Japanese sentences such as /ohayo-gozaimasu/ ("good morning"), /kon'nichiwa/ ("good afternoon"), /kombanwa/ ("good evening"), and /arigato-gozaimasu/ ("thank you"), the consonants/ha/, /go/, /za/, /su/, /ko/, /ni/, /chi/, and /wa/ were clearly heard as continuous sentences although the synthesized voices had no consonant sounds. The bottom part of Fig. 6.24 shows the sound spectrogram for /ohayo-gozaimasu/. From the evaluation results, it was ascertained that users could easily produce arbitral sentences by tracing the guidelines superimposed on the touchpad.

Next, we carried out an evaluation test using 100 randomly synthesized mean-ingless words composed of four different monosyllables. Although the identification of vowels and voiced consonants, such as semivowels (/ya/, /wa/) and nasals (/na/, /ma/), showed a high score (around 70%), the average identification rate was only 33%. It was clear that the low identification rate was due to a lack of random noise components with relatively long period such as /s/, /h/ and /z/.

After investigating what kinds of noise components should be added to the software and how they could be controlled by the user's finger, we added three noise components that correspond roughly to /h/, /s/ and /ts/, as indicated in Fc1, Fc2, Fc3, and Fc4 of Fig. 6.23. These components were arranged inside of a F1-F2 plane of the touchpad as shown in the left part of Fig. 6.22. By these procedures, a higher recognition rate than that for the conventional method was obtained without the noise components.

By using a touchpad with a pressure sensor and by assigning the detected touching pressure to the pitch frequency, the subject could produce some emotional sounds such as "laughter," "surprise," and "disappointment" by controlling the touch pressure and the tracing speed of the finger. In addition, by connecting this tool with a musical keyboard, the subject could sing songs after a short training period. Since 2013, the software has been commercially available, and users can install it onto a smartphone (iPhone) as an application, as seen in the left of Fig. 6.25 [16]. It is expected that the speech production tool can be applied for people with aphasia, especially Broca aphasia and also for one of new musical instruments that can produce songs by touching and tracing a fingertip on display of the smartphone as seen in the right part of the figure.

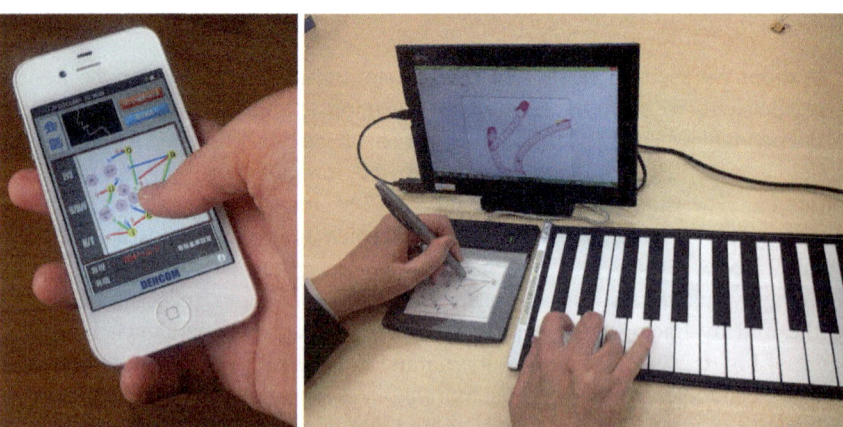

Fig. 6.25 *Left* An application for a smartphone (iPhone 4s), *right* application for a musical instrument connected to a musical keyboard

6.5   Evaluation of Aids and Treatment for Speech Organ Disorders

6.5.1   Vocal Sounds of a Person with a Cleft Palate

As previously described, mynah birds are a kind of mimicking bird that can produce humanlike vocal sounds, despite the fact that the shape of their vocal organs differs greatly from those of a human being. However, in human cases, an abnormal formation of the oral cavity can sometimes lead to dysphonia. Cleft lips or clef palates are birth defects characterized by malformation of the oral cavity. Especially in the case of both a cleft palate and a cleft lip, and functionally imperfect closure of the nose and throat, the sound source waves of the vocal folds are transmitted into the nasal cavity as well as the oral cavity, resulted in a nasal-sounding voice (Fig. 6.26a).

Research on functionally imperfect closure of the nose and throat as well as pronunciation problems resulting from a cleft palate has been conducted in various ways from both the physiological and technological points of view. However, all research findings reported to date have been only qualitative in nature; objective methods of measurement that directly correspond to the degree of auditory dysphonia have yet to be established. As a result, clinical diagnosis and evaluation of dysphonia actually depends on auditory judgments, including such cases as a speech aid (Fig. 6.26b) to prevent functionally imperfect closure of the nose and throat.

Through the use of acoustic techniques, Imai and his associates developed a quantitative method to evaluate vocal sound nasalization resulting from functionally imperfect closure of the nose and throat. When this quantitative method was evaluated by means of both an LPC method and homomorphic analysis, it was found that all subjects with a cleft palate exhibited clear differences in speech during the use versus the non-use of a speech aid, especially in the case of the vowel /i/. The homomorphic method is basically a form of signal processing used for the detection of zero points that can detect to a certain degree leaking sounds resulting from imperfect closure of the nose and throat.

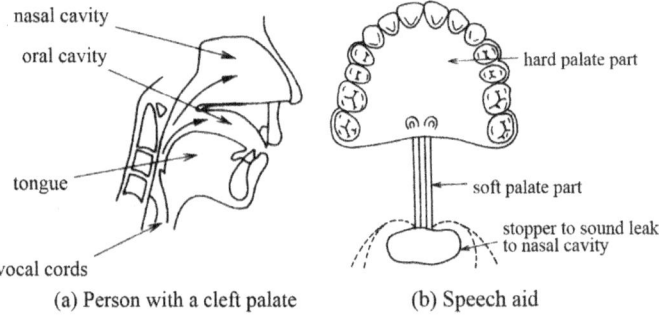

(a) Person with a cleft palate   (b) Speech aid

Fig. 6.26 a Cross section of vocal tract of a person with a cleft palate, **b** speech aid

vowel /i/ of Person with a cleft palate

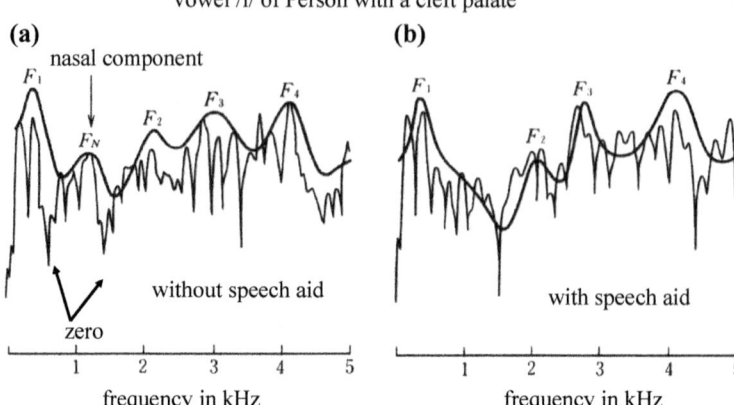

Fig. 6.27 Spectrum of vowel /i/ of a person with a cleft palate: **a** without speech aid, **b** with speech aid

Figure 6.27 shows the spectral pattern of the sound /i/, when vocalized by a person with a cleft palate. Without the speech aid (Fig. 6.27a), two zero points appear between the first and second formants. In addition, a peak (F_N) of the spectrum envelop appear to form a counterpart to them. However, without the speech aid, the peak (F_N) was barely observable. In a hearing experiment on synthetic vocal sounds in which the F_N elements were gradually increased, a very high correlation existed between the scale of F_N for the vowel /i/ and the subjective evaluation of the degree of nasality (Fig. 6.27b).

It is clear that a cleft palate leads to an increase in the sounds passing through the nose—a factor that has an effect on the nasalization of sounds, the extent of which can be evaluated by use of the F_N scale for the vowel /i/ [17]. Such examples confirm that signal processing is of great use for the evaluation and the design of speech aids that can be adapted for each individual with a cleft palate. Next, another form of dysphonia resulting from changes in the shape of a mouth will be described.

6.5.2   Vocal Sounds of a Person with a Reversed Occlusion

Reversed occlusion is one type of malocclusion or misaligned bite, which requires surgery when it has been determined that orthodontic treatment is not sufficient to enable a person to bite properly due to their having a major skeletal problem. Often, these people do not only suffer from an aesthetic and masticatory defect, but also experience articulation disorders. In many cases, articulation disorders appear in consonants, such as fricatives and explosives. Therefore, clinically speaking, it is extremely useful to quantitatively evaluate the extent of the disorder or the need for changes in treatment.

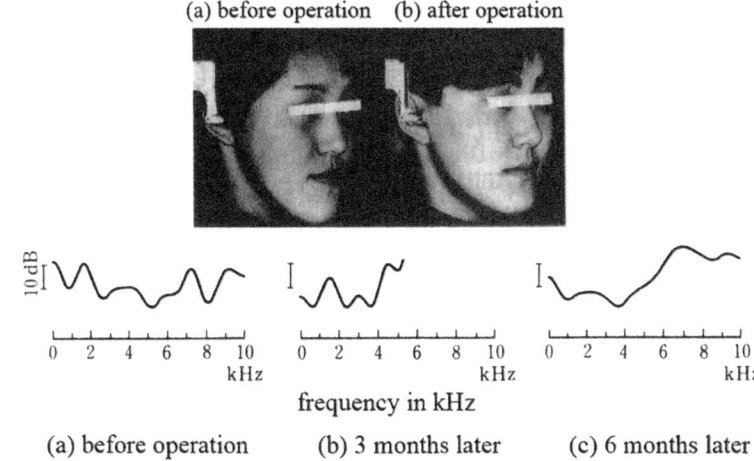

Fig. 6.28   Changes in spectral envelope of consonant /s/

As the upper photograph of Fig. 6.28 shows, an esthetic improvement was observed after an operation to correct malocclusion. However, a method to quantitatively evaluate the degree of articulation disorders has yet to be established. In view of this, the author will briefly describe some examinations of the acoustic characteristics of the voiceless fricative sound /s/, as proper vocalization of this sound is often one of the problems associated with articulation disorders that result from skeletal reversed occlusions.

Fourteen subjects were chosen who had been diagnosed with skeletal reversed occlusion for which they had already received surgical orthodontic treatment (5 males, 9 females, with an average age of 20.3). The subjects were asked to vocalize vowel-consonant-vowel (VCV) syllables, including the voiceless fricative sound /s/. Their vocal sounds were recorded during three different periods: just before undergoing surgery, three months after surgery and finally, six months after surgery.

The lower graphs of Fig. 6.28 show an example of the results of the vocal sound analysis. An increase in the sound pressure level in the high-frequency zone and a decrease in that in the low-frequency zone were clearly observed following the operation. This change is more obvious in the figure (c) than in the figure (b) because the former shows the results obtained six months after surgery, while the latter shows the results obtained three months following the operation [18].

As the articulatory organs that human beings have developed are so finely constructed, even a slight change in the shape of the mouth can cause articulation disorders. Thus, in the area of sound assistive technology, there is clearly a greater need to address the complicated difficulties arising from such disorders.

6.6   The Future of Speech Production Aids

In cases where the larynx has been removed or the voice tract is partially deformed, either speech substitution or surgery is necessary in order to produce vocal sounds with a high degree of naturalness. Recent research on signal processing, however, has also yielded significant progress that is applicable to the synthesizing of vocal sounds. In particular, research is now being directed towards a more natural conversion of synthesized vocal sounds that takes full advantage of the recent advances in signal processing technology.

In particular, it is effective to utilize vocal sound synthesizing devices for those whose speech is extremely unclear because of cerebral paralysis, etc. A famous astrophysicist, Professor Stephen Hawking, skillfully operates a device to synthesize vocal sounds by making use of his surviving motor functions. By watching musicians play musical instruments as if they are a part of their body, one can imagine the future possibility of people handling a voice synthesis device like a musical instrument. In order to achieve this goal, research should be more strongly pursued with the goal of designing an interface for a voice synthesis device that is better suited to the characteristics of a brain function regarding language.

References

1. N. Kobayashi, Assessment of articulation by auditory impression. Jap. J. Logopedics Phoniatrics **41**, 142–146 (2000)
2. L.D. Lowry, Artivial Larynses: A review a development of a prototype self-contained intra-oral artificial. The Laryngoscope **91**, 1332–1355 (1981)
3. M.I. Singer, E.D. Blom, An endoscopic technique for restoration of voice after larygectomy. Ann. Otol. Rhinol. Laryngol. **89**, 529–533 (1980)
4. T. Hirahara, T. Ifukube, C. Yoshimoto, An articulation model of a talking bird "Mynah": an analysis of mimic voices pronounced in He-O$_2$ atmosphere. J. Acounstical Soc. Jap. **38**(6), 321–329 (1982)
5. N. Uemi, T. Ifukube, M. Takahashi, J. Matsushima, *Design of a new Electrolarynx having a Pitch Control function, Robot and Human Communication, 1994. RO-MAN '94 Nagoya, Proceedings*. 3rd IEEE International Workshop on Speech (1994), pp. 198–203 (in Japanese)
6. T. Ifukube, M. Hashiba, J. Matsushima, A role of "waveform fluctuation" on the neutrality of vowels. J. Acoust. Soc. Jap. **47**(12), 903–910 (1991)
7. R.G. McRae, H.R. Pillsbury, A modified intraoral electrolarynx. Arch Otolaryngol **105**, 360–361 (1979)
8. UltraVoice, http://www.ultravoice.com/ [2016.12]
9. E.A. Goldstein, J.T. Heaton, J.B. Kobler, G.B. Stanley, R.E. Hillman, Design and Implementation of a Hands-Free Electrolarynx Device Controlled by Neck Strap Muscle Electromyographic Activity. IEEE Trans. Biomed. Eng. **51**(2) (2004)
10. C. Painter, T. Kaiser, J.M. Fredrickson, R. Karzon, Human speech development for an implantable artificial larynx. Ann. Otol. Rhinol. Laryngol. **96**(5) (1987)
11. M. Hashiba, Y. Sugai, T. Izumi, S. Ino, T. Ifukube, Development of a wearable electro-larynx for laryngectomees and its evaluation, in *Proceedings of the 29th Annual International Conference of the IEEE EMBS* (2007), pp. 5267–5270

12. H. Min, M. Takahashi, N. Nishizawa, S. Nisizawa, T. Ifukube, Y. Inuyama, Control of the Sternohyoid Muscle for an Electrolarynx. Jap. J. Med. Electron. Biol. Eng. **32**(4), 69–77 (1994)
13. T. Ifukube, Mynah bird, parakeet and ultra-ventriloquism: to solve the mystery of the speech production. Acoust. Soc. Jap. **56**(9), 657–662 (2000, in Japanese)
14. K. Yabu, S. Aomura, T. Ifukube, Proposal of a speech-synthesis interface for speech disorders by using a pointing device human interface. Trans. Hum. Interface Soc. **11**(4), 437–447 (2009). (in Japanese)
15. P. Delattre, F.S. Cooper, A.M. Liberman, L. Gerstman, Acoustic Loci and transitional cues for consonants. J. Acoust. Soc. Am. **26**(1), p137 (1954)
16. M. Hashiba, K. Yabu, M. Takase, Y. Sugai, T. Ifukube, Development of a real-time speech synthesis application software for smartphone controlled by finger positions and movements, Human Interface Society, in *Proceedings of the Human Interface Symposium* (2013), pp. 507–510 (in Japanese)
17. T. Imai, S. Nakamura, T. Hirahara, T. Ifukube, Quantitative evaluation of the nasality in cleft palate speech. J. Acounst. Soc. Jap. **41**(2), 69–76 (1985). (in Japanese)
18. T. Yamamoto, T. Imai, S. Nakamura, T. Ifukube, Acoustic characteristics of the /s/ fricative in the skeletal class III malocclusion. J. Acounst. Soc. Jap. **47**(9), 626–631 (1991). (in Japanese)

Chapter 7
Sound Information Aiding for the Visually Impaired

Abstract In this chapter, the author mainly discusses two assistive tools for the visually impaired. One of these is a kind of screen reader that we call the "Tactile Jog-dial". It converts verbal information such as text into speech signals for which the speech speed can be controlled by blind users, while displaying non-verbal information such as rich text to the tactile sense of a fingertip. This is based on the experimental result that most blind people can recognize spoken language at around three times the speech speed than standard. The other are mobility aid devices that detect environmental information and display it to the auditory sense of the blind using sounds so that the blind can recognize the environment, especially obstacles, surrounding them. These were modeled after the echolocation function of bats and also an ability of obstacle sense that the blind acquire. Moreover, he talks about a method of controlling the balance function by sound localization.

7.1 Character-to-Speech Conversion Systems

7.1.1 Text-to-Speech Conversions

Computers that have a text-to-speech (TTS) converter and an optical character reader (OCR) have enabled blind users to read and write documents in digital format. Before personal computers appeared and became accessible to blind users, the blind needed to depend on Braille or audiotapes translated by sighted people to obtain information. Translation usually takes quite a lot of time, in addition, the physical size of a book written in Braille is very large. In the case of audiotapes, it also takes a lot of time to find the necessary information as the blind have to navigate through audiotapes solely by using the forward and backward functions of an audiotape player [1].

In the 1980s, blind people could read digital documents stored on CD-ROM, or even read a printed document using a scanner and OCR software [2]. The Kurzweil

© Springer International Publishing AG 2017
T. Ifukube, *Sound-Based Assistive Technology*,
DOI 10.1007/978-3-319-47997-2_7

system, manufactured by the Kurzweil Corporation, is a system that converts English sentences into vocal sounds. This device is capable of understanding the meaning of English sentences that can be converted into vocal sounds with a natural intonation. Furthermore, some such devices are able to change the speed of synthetic sounds and choose male or female voices.

In fiscal 1980, the Japanese government embarked on a five-year plan to develop a system of converting written Japanese into vocal sounds modeled after the Kurzweil system. This special research was titled "Research regarding devices to enable the visually impaired to read Japanese by means of sound conversion [3]." In 1982, using minicomputers, a prototype was made of an auto-scanning system that converted written Japanese into vocal sounds, and in the last fiscal year of the plan (1985), a working model was made of a similar approach that utilized a multiple-computer system to vocalize written Japanese.

At that time, the software's built-in dictionary contained a total of 2041 Japanese characters and 8600 rules governing language use. An evaluation of this system was conducted to determine the accuracy of input words, recognition of meaning and vocalized speech output. For the sample of characters in the recognition experiment, three letter styles that included 996 characters were selected from a typical elementary school textbook. As the resultant accuracy rate was between 95 and 97%, this offered sufficient proof of the practicality of algorithms for character recognition.

However, there were some opinions voiced regarding the quality of the vocal sound output for the system that converts Japanese characters, including Chinese characters called *kanji* as mentioned in Chap. 5, into vocal sounds. For the output system, which consisted of the prompt reading of vocal sounds without intonation or accent, the opinions were as follows:

(1) When congenitally blind subjects are under 9 years of age, it is important for them to learn the shapes of the characters prior to learning their accents.
(2) As reading is categorized as an auditory language, as opposed to a visual one, it is therefore natural to consider such essential factors as accent, intonation and speed.
(3) Some other methods such as the Optacon (see Chap. 4) are better, as a person can touch the letters or shapes and hear the corresponding sounds. In such a situation, correct intonation would not be necessary.

As each one of these opinions is related to the input of text through computers, the people holding such viewpoints naturally have a strong interest in being able to use computers for reading. It has also become clear that for the blind who have access to computers, one of their greatest concerns is to be able to somehow output Japanese in the JIS (Japanese Industrial Standards) code as part of some form of a multipurpose interface system.

7.1.2 Screen Readers

Word processors for blind people were first developed in the early 1980s in Japan. A few years later, screen readers for MS-DOS® were developed. However, there are some fundamental issues regarding accessing visual documents using existing screen readers. For example, blind users utilize cursor keys for exploration, while sighted users use their eye movements to guide their exploration. Eye movements can be flexibly controlled to read the information slowly or rapidly, based on the needs. A cursor-based exploration only allows reading aloud between lines and characters by moving the cursor keys. Operating systems (OSs) based strongly on graphical user interfaces (GUIs) sense have spread in personal computer environments. However, they are inherently difficult for blind people to operate as they cannot use visual information.

With this problem in mind, Watanabe and his colleagues proposed a screen reader program as a support tool for GUI operation by blind people [4]. The proposed software transmits the information on a screen through synthesized speech for the benefit of blind users. It does not require the use of pointing devices; all computer operations can be done from the keyboard. The screen reader builds an environment in which blind people can use applications like word processors in Microsoft Windows©. In view of the above, they developed a prototype screen reader in 1996 called the "95Reader©." Their screen reader for Windows consists of several pieces of software, as shown in Fig. 7.1a. To improve software for blind people, an MS-DOS front-end processor with speech output was revised to work in the Windows environment. This program not only converts Japanese syllables *kana* into *kanji*, but also outputs them as understandable verbal expressions.

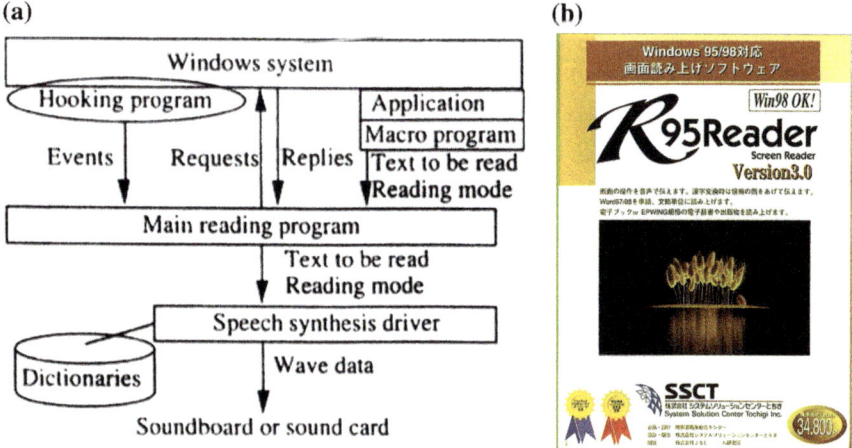

Fig. 7.1 **a** Block diagram representation of the screen reader "95Reader," **b** screen reader packaging case

Based on evaluation tests by 52 blind users, the prototype screen reader was revised. For example, a reading mode was added as a parameter to be sent from the main reading program to the speech synthesis driver, enabling the speed, the pitch, the length of pauses at punctuation, the accent and the speaker (male, female, or computer) to be chosen by the user. The revised screen reader was put into practical use in 1997 and has been widely used throughout Japan. It was priced at $300 (U.S.).

The screen reader has been improved to make it compatible with Windows 98© and Windows 2000©. Figure 7.1b shows the packaging case for the screen reader named "95Reader." Almost 80% of blind users in Japan used this screen reader. Nowadays, if they can utilize a network application such as email, people—including blind workers—can receive documents using spoken text.

In the mid-1990s, GUI-based OSs that represented information visually in the form of images and movies spread rapidly in personal computer environments. Such GUI systems brought an interface that drastically improves the ease of creating visual documents. However, the blind need to keep listening to spoken words until they hear the exact sentence they need. In addition, speech speed control is only statically available using a menu or keyboard command. This means that they cannot dynamically slow down or speed up the rate while listening to the spoken documents. Furthermore, without layout information they cannot recognize when a new chapter or a new paragraph is started. Background colors on a Web page are often used to group information. Without knowing about such background color information, it is hard for them to notice the organization of groups. "Rich text" features—such as font sizes and bold attributes—are used to emphasize words or sentences in a document. Without knowing about such emphasized information, they may easily miss important points of a document.

Furthermore, experienced blind computer users have complained that the default and the fastest speech speeds are too slow. Therefore, blind people would greatly benefit from a breakthrough in technology to improve the non-visual interface and to explore visual information non-visually as quickly and effectively as possible, and to represent visual effects non-visually using as many modalities as possible to retain the benefits for blind people of using computers.

7.1.3   Tactile Jog-Dial (TAJODA) Interface [5]

With this problem in mind, Asakawa, who is a blind researcher, her co-researchers and the author proposed a "tactile jog-dial" (TAJODA) that can control the speech speed of a screen reader while displaying rich text features to the tactile sense of a fingertip. Figure 7.2 shows a block diagram of the TAJODA. After a long survey and discussion, a jog-dial was selected as a substitute for the cursor for editing and exploring multimedia data. The jog-dial interactive device makes possible intuitive operations on speech information. The speed at which the dial is spun affects the speech speed, which allows for dynamic and variable speech speed control. Selected rich text information such as font family, font size, and bold script were then designed

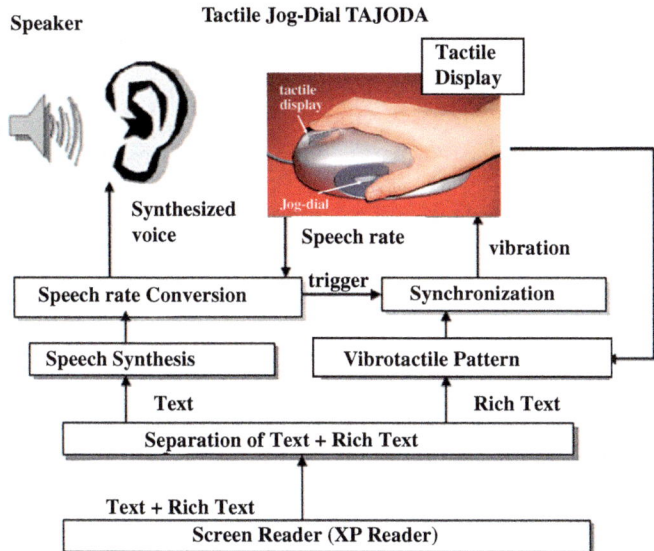

Fig. 7.2 Block diagram of the TAJODA interface. The speech rate can be controlled by the thumb. Rich text is displayed as vibro-tactile patterns to the tip of the index finger

to be presented on the tactile device. Both speech and tactile information can be synchronized with the dial movements.

7.1.3.1 Maximum Speech Speed

In designing the TAJODA, first tests on the maximum listening speed were performed on blind users to investigate the dependence of their correct recall rate on the speech presentation speed (speech rate) in morae per minute. The experiment consisted of subjective evaluations based on subjects' opinions and an objective evaluation based on their recall accuracy (correct recall rate) in hearing short sentences. From the results of the recall accuracy obtained for about 20 blind subjects, as shown in Fig. 7.3, it was found that the speech rate at 50% correct recall rate was about 1400 morae per minute for experienced users (subject A, B, D in the figure). This speech rate shows 2.6 times faster than the average speech rate (550 morae per minute) for the sighted [6]. The novice users' average speech rate at 50% correct recall rate was around 1.5 times faster than the default rate for the TTS system (500 *morae* per minute). Furthermore, as mentioned in Chap. 2, Imai and his colleagues proposed "adaptive SRC" algorithm that retains only speech parts that are likely to have a significant effect on listening comprehension. They reported that the visually impaired people can comprehend the speech voices much easier than the conventional SRC [2.13]. The results were incorporated into the variable speech speed control interface for the TAJODA.

Fig. 7.3 Correct recall rate as a function of speech presentation speed in morae/min

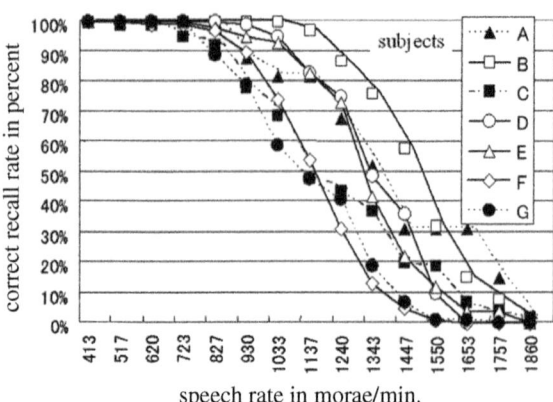

speech rate in morae/min.

In the next step, based on psychophysical experiments, we investigated what kind of rich text information should be presented as tactile patterns. To select the target information, we collected statistical data on frequently used rich text information in existing documents. Seven attributes, as indicated in the right part of Fig. 7.4, were selected from frequently used rich text and layout information, based on document analysis and previous research work. The tactile display that we used was a piezoelectric vibrator driving 32 pins, arranged in two columns of 16 pins each, as shown in left of the figure. As a result, we determined some parameters in the prototype model that can control speech speed in real time by the use of a finger while touching the rich text with another finger.

7.1.3.2 Evaluation of TAJODA

We evaluated the efficiencies of the TAJODA from the viewpoint of short sentence recognition, consisting of normal text and three kinds of rich text information. The experiments compared the following four methods, labeled SCR, TAC, JOG, and TAJODA.

(1) Screen Reader (SCR) method: SCR simulates similar functions with a screen-reader-based reading and exploration method. The cursor up and down keys were used to explore the test data. The rich text information was presented in the form of speech, such as "bold" (stating bold and "18 point". These spoken cues were inserted into the test data as one segment, just like the usual screen reader rich text reading method.

(2) Screen reader and Tactile (TAC) method: TAC presented the rich text information in the form of tactile cues, corresponding to each segment's attribute.

(3) JOG-dial (JOG) method: JOG used the shuttle/jog dial to explore the test data, and the rich text information was presented in the same way as for SCR.

(4) Tactile and Jog-Dial (TAJODA) method: TAJODA used all of the functions of the experimental TAJODA software and devices.

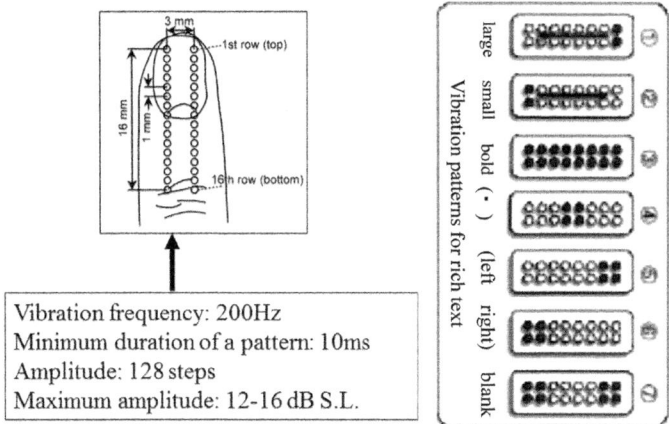

Fig. 7.4 *Right* Seven attributes of rich text answer as a function of speech speed. *Left* A tactile display for TAJODA

In each test set for each method, blind users were asked to say which segment has what attribute. They were told that three kinds of attributes could appear, but they were not told how many attributes were included in any of the test sets.

Figure 7.5a shows a box-and-whisker plot of the times for the four methods for all subjects. There was no significant difference between SCR and TAC. JOG was significantly faster than SCR and TAC. TAJODA was significantly faster than the other three methods. TAJODA (32.5 s on average) was 2.4 times faster than SCR (78.0 s on average).

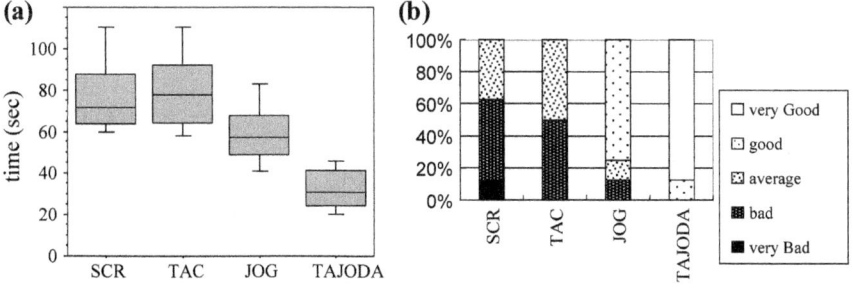

Fig. 7.5 **a** Result of time trial comparison test, **b** comparison of subjective evaluation of four methods

An evaluation test of the system showed that the sense of touch could be effectively used to understand speech information, and that using tactile patterns could speed up the non-visual exploration time. This allowed users to actively and flexibly navigate through the speech information by dynamically controlling the speech speed. Actually, the use of the jog-dial was well accepted by all of the subjects from a viewpoint of subjective evaluation, as shown in Fig. 7.5b. The tactile cues also helped them to look for particular information in the speech information that was only available through a linear search.

The blind researcher ASAKAWA for the TAJODA project said: "The tactile cues act like guideposts on a road. The road is like a string of speech information and the guideposts are the tactile cues. It can easily be imagined how much and how well the non-visual exploration time can be improved by the use of tactile cues."

7.1.4   Auditory Displays Using Sound Localization and Lateralization

7.1.4.1   Displays Using Lateralization

Various approaches have been used to convert letters and images into acoustic stimulation. Most of them aim at enabling a blind person to recognize letters and images using a sound's lateralization as well as pitch information perceived through a binaural headphone. In these approaches, the perceived position of a sound image inside the subject's brain and the pitch frequency corresponds to the horizontal axis and the vertical axis of text and images, respectively. "Lexiphone" and an auditory display designed by Itoh and his co-researchers are typical examples [7, 8]. The latter method involves placing the sound reference point inside the brain, represented by 0, exactly halfway between the two headphones, as shown in Fig. 7.6a.

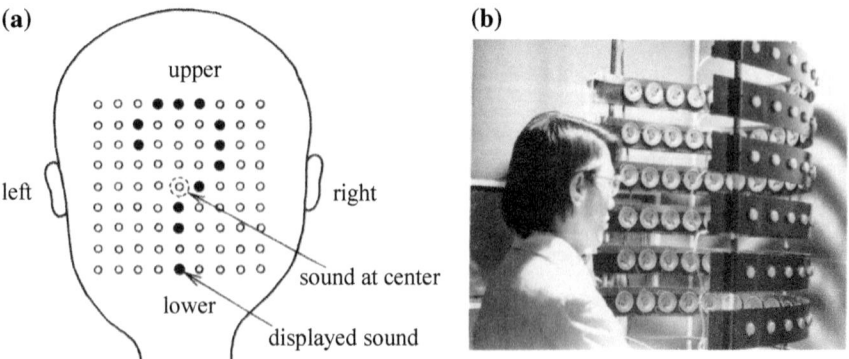

Fig. 7.6 **a** Arrangement of sound lateralization, **b** speaker matrix (6 columns × 16 rows)

In Itoh's approach, the equal temperament scale's lowest sound was fixed at 262 Hz with a frequency ratio of $f_n : f_{n+1} = 1 : 1.26 (n = 1 - 9)$ for each emitted sound. They reported that a correct recognition rate of about 80% for 68 *kana* was achieved when the sounds were emitted at 50 ms/character.

The author and his co-researchers developed a sound location system by placing a large number of compact speakers around a subject's head, including the ears and auricles as shown in Fig. 7.6b. Therefore, it may become possible not only to recognize horizontal sound images, but also to recognize the vertical sound images and plural images that were formerly difficult to recognize [9]. The layout of a speaker matrix consisted of 96 small speakers with a diameter of 3 cm arranged into 6 columns (10 cm apart), and 16 rows (8 cm apart) in a semicircle with a radius of 30 cm. To maintain equalization, both the dispersion of the frequency characteristics and the sound pressure range of each speaker were set as high as possible. Using this experimental system, it was possible to choose certain speakers on an arbitrary basis to generate sounds of an optional intensity, frequency and duration. The author will mention in the following session how correctly subjects can perceive the vertical location of the sound sources and can identify the characters displayed on the speaker matrix as localized sound images. Incidentally, the details of regarding to localized sound images are precisely mentioned in a book "Spatial Hearing" [10].

7.1.4.2 Display Using Localization

The subjects were five males aged from 22 to 23, included four males with normal hearing and eyesight, and one male whose hearing in the right ear was normal but who had lost 85 dB in his left ear. The subjects' heads were fixed almost in the center of the speaker matrix and the character identification experiment was started after four or five trial tests had been conducted.

The quantity of information conveyed through the auditory sense is limited by the spatial and temporal resolution of the senses. Therefore, the intervals used to arrange the speakers in the speaker matrix were set slightly lower than that of the spatial resolution threshold. However, as the vertical location characteristics and some sound sources of the auditory sense were mostly unknown at that time, these characteristics underwent careful examination in this basic experiment.

First, by using only the speakers of the six central rows, the author randomly chose one of those speakers to generate bursts of white noise. After hearing the noise, the subjects were asked if the sound location appeared to be at the same height as their ears or at a point that was higher or lower than their ears. It was ascertained that as the vertical position of the sound source moved higher, the subjective height of the sound perceived by the subjects was also proportionally higher as shown in Fig. 7.7a. It was found that the average spatial vertical resolution was around 15° in this experiment.

However, when ear pads made of acrylic board or a piece of silicon with a hole in the center were worn over a subject's ears and the same experiment was constructed, it became clear that sounds generated from below a subject's ears also

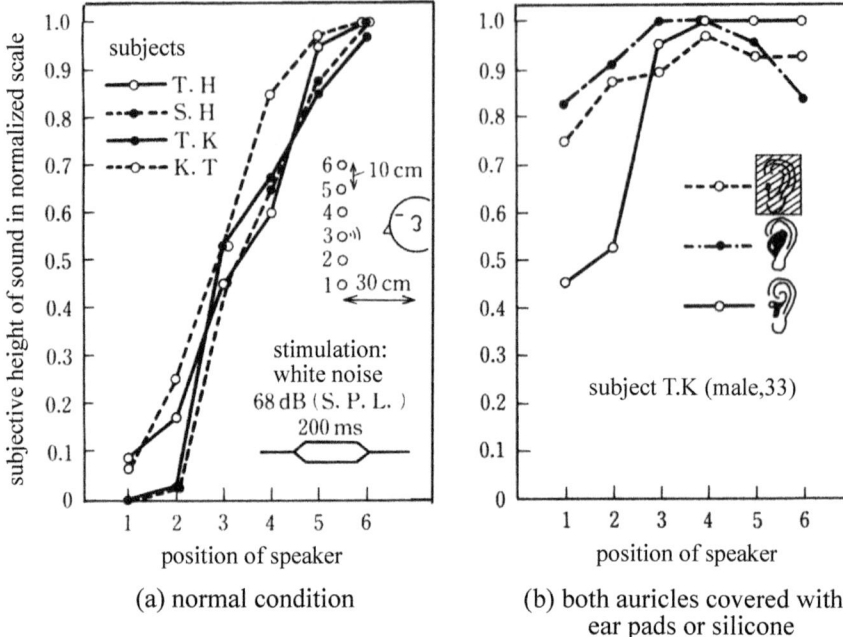

Fig. 7.7 a Vertical sound localization in normal condition, **b** vertical sound location when the subject's auricles are covered with acrylic board or a piece of silicon

appeared to be emitted from a location above the ears (Fig. 7.7b). These results supported the conventional hypothesis that the auricles, especially the *tragus* shown in a solid line of the figure, play an important role in the vertical sense of sound [11].

Next, in the case of multiple sound sources, one possible problem was that the perception of the precise location of the sound might shift, thereby giving rise to the chance of error in the recognition of characters and images. In view of this possibility, the author and his co-researchers examined how the sound perception of pure tone output from the central speaker shifts when white noise is generated from the right. From the result, it was ascertained that the location of the sound image shifted from the center to the left according to the intensity of white noise; for example, the shift was about 6° from the center to the left when a noise at 59 dB [S.L.] was emitted from a speaker on the rightmost. The most logical explanation is that the pure tone entering into the left ear was weakened by the masking effect of the white noise in the left ear. Thus, as the presentation of several sources of sound results in the subjective location of the sound source shifting from its actual location, it would appear that each sound source must be generated sequentially in order to facilitate pattern recognition [12].

Bearing this in mind, we investigated how the character images emitted from the speaker matrix depend on the period of the sound presentation of each character

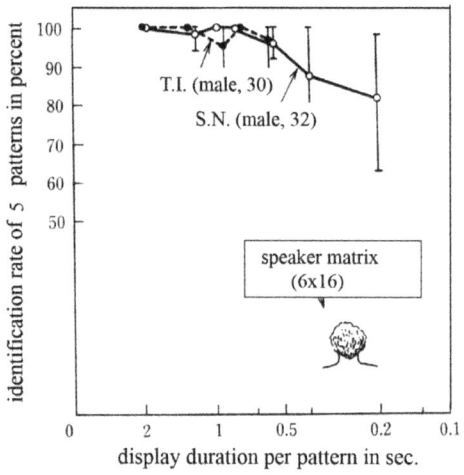
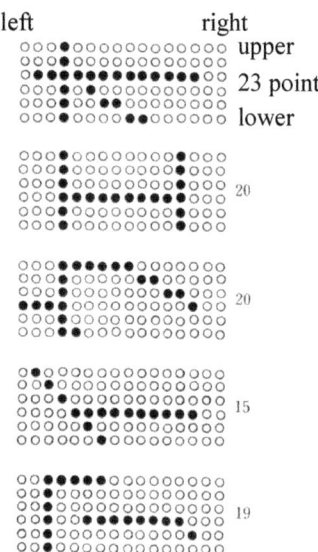

Fig. 7.8 *Right* Five patterns (Japanese *kana*) displayed on a speaker matrix, *left* identification rate of the five patterns as a function of displayed duration

pattern. Five Japanese *kana* syllables were used as the character patterns as shown in the right side of Fig. 7.8, and a sound stimulation was generated from the speakers, represented by the black spots in the figure. During this examination, two different orders of presentation were used: (1) Beginning from the highest position at the left side and finishing with the lowest position at the right side, and (2) Tracing the writing order of each of the *kana* syllables [13].

Following a training period of approximately 30 min, the identification rate became close to 100% for a presentation period of 1 s. However, the identification rate gradually decreased as the presentation period was shortened. On the other hand, despite the same presentation period, when the order of presentation consisted of tracing the writing order of the syllables, a generally higher identification rate was obtained. A possible explanation for this result is that it is easier to combine the writing order with syllabic patterns that have previously been memorized. The identification rate for subjects with an impairment in only one ear was roughly 60%.

It is difficult to determine any clear degree of practical application from the above findings as the characteristics of syllable recognition are widely influenced by such complicated factors as learning effects and individual differences among subjects. It would be worthwhile to examine ways to improve the recognition of more complicated patterns, colors and 3-D objects as the quantity of the information conveyed could still be increased by changing the frequency and tone of each sound source. Furthermore, considering that the blind mainly depend on the location of a sound source to recognize obstacles that they cannot touch, this method could at least be utilized as a training device for obstacle recognition and, in the near future, as a 3-D sound display for virtual reality systems.

7.2 Mobility Aids for the Blind

The greatest everyday inconvenience for the blind is the difficulty in walking from place to place. For the blind, walking can be divided into two aspects: orientation and mobility. It is easy to imagine how important it is for the blind to somehow acquire this orientation ability—a role normally played by the visual sense. In order to recognize their surroundings, it is necessary for the blind to either attempt to compensate for their lack of visual sense through the use of their remaining senses or to utilize assistive devices to perform certain tasks that their disability makes it difficult or impossible for them to perform.

In the 1970s, various studies were conducted on devices to compensate for visual disabilities, including electronic-based systems, such as a Laser cane, which help the blind to walk [14, 15]. The purpose of electronic-assistive devices for walking is not to replace white canes or guide dogs, but rather to provide further information that the former methods cannot offer. Some such devices were commercialized, with the two most readily available and widely used being the "Mowat Sensor" and the "Sonic-guide", which were first put on the market more than 40 years ago. Despite the remarkable progress in electronic technology in recent years, more efficient, user-friendly and affordable electronic mobility aids have yet to appear on the market.

One of the reasons for this delay is that research on electronic mobility aids is mostly limited to the technical field led by optical researchers. Specifically, in order to make new devices practical for rehabilitation and education, improvements must be made in the function of hardware, coupled with completion of the necessary research on such factors as the preparation of training manuals and clearly established methods of instruction. It is therefore the author's intention to introduce and emphasize the need for more basic research as well as the necessity to improve both hardware and software for electronic mobility aids.

7.2.1 Mowat Sensor and Sonic Guide

Both the Mowat Sensor and the Sonic Guide employ a method of emitting ultrasound from a user to her or his surroundings in order to detect obstacles by means of reverberations or echoes, although there are some differences in their relative efficiency and principles of detection [16].

The Mowat Sensor is a small rectangular device (15 cm × 5 cm × 2.5 cm, 175 g) developed by Mowat in 1972, as shown in Fig. 7.9a. This device emits ultrasound pulses to detect obstacles based on reverberation. Its range of direction

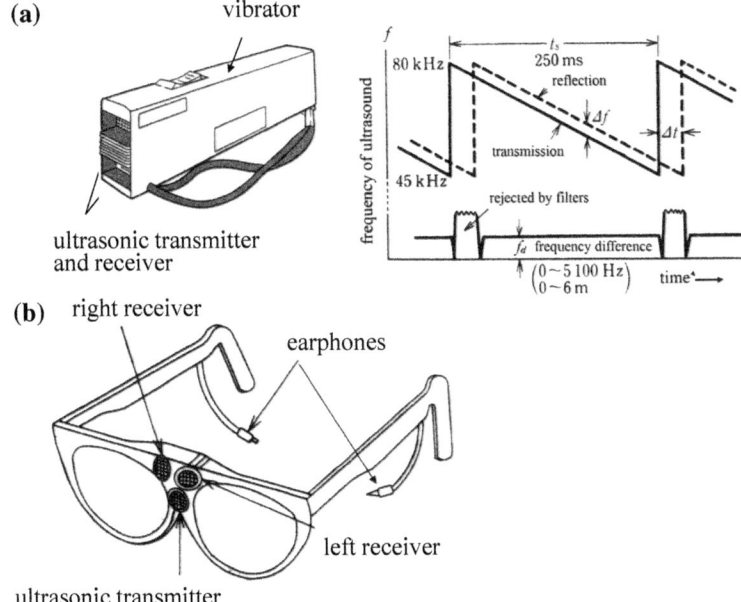

(a) vibrator

ultrasonic transmitter
and receiver

(b) right receiver

earphones

left receiver

ultrasonic transmitter

Fig. 7.9 *Left* Appearances of Mowat Sensor (**a**) and Sonic Guide (**b**), *right* principle of distance detection for Sonic Guide

from the front of the user is 15° horizontally and 30° vertically. To inform the user of the location of an obstacle, the entire device vibrates in accordance with the user's distance from the obstacle. As the user approaches the obstacle or object, the vibrational frequency increases proportionally. For example, at a distance of 1 m, the device vibrates at about 40 Hz, whereas at a distance of only 25 cm, the frequency increases to 130 Hz. The Mowat Sensor makes two detection zones available to the user; one within a 1-m radius and the other within a 4-m radius, each of which can be selected by turning a switch. When there is more than one obstacle, the device only reacts to the one closest to the user. As the device is equipped with an earphone terminal, it is also possible to judge distance according to the sound pitch that corresponds to the vibrational frequency.

Just as people with normal vision use a flashlight, the Mowat Sensor can be held in the hand as a scanner to determine the direction of obstacles. However, as the width of the ultrasound beam is narrow, it is necessary to become quite adept in this requisite scanning technique to gain sufficient information about one's surroundings. Another drawback of this device is that it causes both hands to be occupied as the user must learn to use it together with a white cane. According to data obtained in 1988, 85 such devices were in use in Japan.

The Sonic Guide was developed under the leadership of Kay [17] was the most widely used of such devices. It is fundamentally based on the principle known as echolocation, a function that bats use to detect obstacles or capture prey by use of ultrasound.

Most bats emit ultrasound with a fundamental frequency of between 40 and 80 kHz. There are two types of such bats: constant frequency and frequency modulated (CF-FM) bats, which change a sound's pitch after emitting sounds at a fixed frequency; and FM bats, whose sound frequency decreases from the beginning of its emission. The Sonic Guide emits ultrasound similar to that used by FM bats. FM bats can emit sounds several times per second, whereas the Sonic Guide emits sounds at a slower rate.

As the Fig. 7.9b and the right side figure show, an ultrasound emitter that is fixed on the upper ridge of a pair of glasses just above the nose, emits successive FM ultrasonic bursts at frequencies from 80 to 45 kHz in cycles of 250 ms. Two ultrasonic wave receivers fixed on either sides of the emitter receive the reflected waves three dimensionally. The frequency of the emitted ultrasound drops within 0.25 s, followed by a sudden increase.

In this way, the two receivers are able to detect the reflected sounds from obstacles. Beats are generated by simply adding delayed signals and sending them. When these beats are transformed into sounds, they are heard as beat sounds. The Δt in the right figure signifies the time delay of the reflected sound; indicating the distance from the user to an obstacle. The closer the distance to the obstacle, the smaller the value of Δt. Therefore, the frequency difference Δf between the emitted and reflected sound decreases. Consequently, the user hears a lower sound.

The two ultrasonic receivers are mounted on the glasses at an outward-facing angle of 15°, so as to create a difference in sound pressure between them that corresponds to an object's direction. In this way, a user is able to determine the direction of obstacles within an angle of 40° on each side. However, some problems with the Sonic Guide are that the signal-to-noise (S/N) ratio decreases when another object is behind a given obstacle. Moreover, ghost images occur when more than one obstacle is perceived. Another limitation of this device occurs when similar obstacles are located to the right and left side of a user. In this case, the user receives the mistaken impression that an obstacle exists where there is in fact no such obstacle.

In terms of perception, the sounds heard in this case are actually the opposite of the sensation that the blind usually experience; that is to say, the closer the obstacle is to the user, the higher the sound, while the further the obstacle, the lower the sound. Furthermore, many blind people could not afford to purchase the device as it was expensive. In fact, data obtained in 1988 indicated that only 130 Sonic-guides were in use throughout Japan, and only 2000 were in use throughout the world.

Around 1970, under the guidance of the Japanese government, a special project was embarked upon with the goal of creating mobility aids for the blind that addressed the limitations of Sonic Guide glasses. The approach adopted by researchers in this project was to conduct a more in-depth investigation of the basic mechanism of echolocation used by bats as a model for improving the design of such devices. As part of this project, certain findings were obtained from the participation of the author, the results of which will be explained next.

7.2.2 Mobility Aids Modeled After the Echolocation System of Bats [18]

Figure 7.10 shows typical spectral patterns for an FM bat and a CF-FM bat. In the hope of providing evidence that an echolocation function similar to that of bats could be applied to the human auditory sense, Sasaki and the author performed both psychological and physical experiments that examined what particular form of ultrasound would be most effective in determining the location of a single target sound.

7.2.2.1 Ultrasonic Eyeglasses Modeled After the CF-FM Bat

The greater horseshoe bat (called *Kikugashira* in Japanese) is one of the most well-known species of CF-FM bats that inhabit Japan [upper photograph of the Fig. 7.10a]. By using such bats as a point of reference, a device was developed that is capable of emitting sounds identical to a bat's ultrasonic waves, thereby enabling

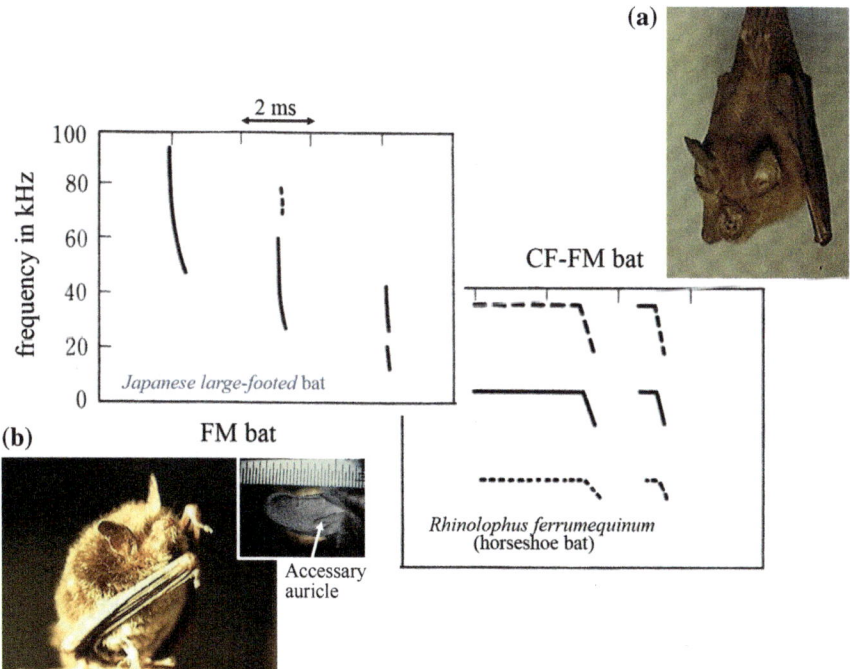

Fig. 7.10 Typical time spectral patterns for ultrasound emitted by bats. **a** Appearance of a CF-FM bat (*greater horseshoe*) and its spectral pattern, **b** that of a FM bat (*Japanese large-footed bat*) and its spectral pattern

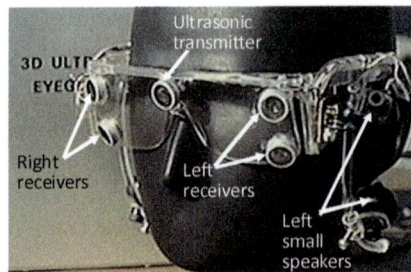
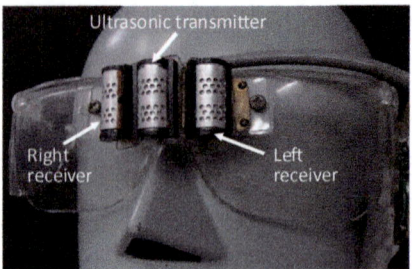

Ultrasonic eyeglasses
modeled after a CF-FM bat

Ultrasonic eyeglasses
modeled after a FM bat

Fig. 7.11 *Left* ultrasonic eyeglasses that display both horizontal and vertical positions of obstacles, *Right* ultrasonic eyeglasses modeled after FM bat

people to hear a sound reflected from an obstacle. Ultrasonic waves emitted from one transmitter of the glasses are reflected from an obstacle, after which they are detected by four receivers fixed at the four corners of the glasses as shown in the left of Fig. 7.11. After converting the waves into sounds that are detectable to the human ear, the reflected sounds are heard through four compact ultrasonic receivers fixed on the upper and lower parts of the front side of both auricles.

From evaluation tests of this device, it was confirmed that the movement of an obstacle or a user causes the sound pitch to increase or decrease due to the Doppler effect. Therefore, this effect enables a user to make an approximate determination of both an obstacle's distance and direction relative to the user. However, in the absence of any movement from either the user or the obstacle, nothing is heard. It is assumed that greater horseshoe bats naturally change the frequency of their ultrasound emissions in order that the reflected sounds always remain the same. This finding might therefore offer one method of introducing an adaptable control system for ultrasonic wave glasses.

The practicality of such mobility aids is premised on their being used in combination with a white cane, as the cane allows for the detection of obstacles that lie one step ahead of the user. In order that this device may also perform a function similar to the white cane, information must be provided regarding the vertical direction of such obstacles. As previously mentioned, it is a known fact that the accuracy and ability for sensing the vertical direction of a sound source depend largely on a person's auricles. As their accessory auricles (*tragi*, see Fig. 7.10b), are abnormally well-developed, it is known that such bats are incapable of flight once their accessory auricles are removed. Consequently, their auricles should provide a hint for the conveyance of vertical obstacle information to the blind.

It has been confirmed that this method can enable users to gain a rough recognition of the vertical direction of obstacles. However, in a more practical sense, one shortcoming of this device was its rather bulky and striking appearance. Nonetheless, the knowledge accrued from this research in how to create a sense of vertical direction did offer significant progress.

7.2.2.2 Ultrasonic Eyeglasses Modeled After the FM Bat

As mentioned earlier, the Japanese large-footed bat (in the lower part of Fig. 7.10b), represents a typical FM bat whose ultrasound emissions undergo a sudden drop in frequency, specifically from 80 to 40 kHz in 1 ms. Based on the assumption that this sudden decrease in frequency may improve a bat's ability to detect obstacles or prey, the author and Sasaki fitted two receivers onto a pair of glasses to develop a device capable of detecting obstacles by emitting sounds identical to those of Japanese large-footed bat.

The appearance and principle chart for this device is shown in the right of Figs. 7.11 and 7.12, respectively. As this mechanism prolongs the reflected sound

Fig. 7.12 Block diagram of ultrasonic eyeglasses modeled after the FM bat

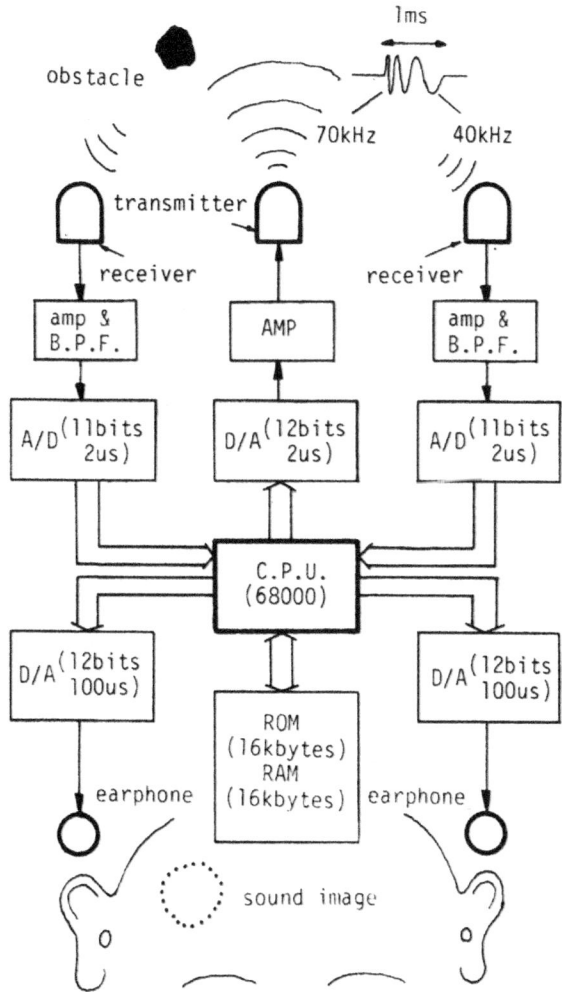

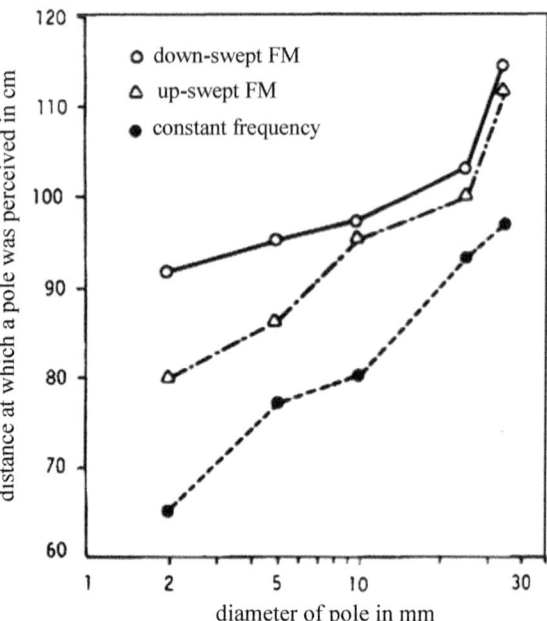

Fig. 7.13 Distance at which a pole was perceived as a function of the diameter of the pole

over time by means of a microcomputer, a frequency of approximately 1/50th is achieved without any loss of wave information. Moreover, the frequency is shifted to a low band whose frequency discrimination score falls within the hearing sensitivity range of human beings. Because past ultrasonic transducers were limited by their strict directionality, it was difficult to distinguish among multiple obstacles. On the other hand, it is a known fact that bats are able to use ultrasonic waves whose directionality is considerable.

We used FM ultrasonic waves with a wide degree of directionality for sound orientation. Some experiments in obstacle detection were conducted, followed by a discussion as to what kinds of ultrasonic sound wave were the most advantageous for echolocation.

In one of experiments, measurements were taken to determine how close a pole (as an obstacle) had to be placed in front of a person, for sounds to be perceived. Figure 7.13 shows the diameter of the pole and the distance from the subject to the pole when sound perception was achieved. The experiment was conducted using both the CF sounds, emitted by the Japanese large-footed bat, as well as FM sounds, whose frequency undergoes a sudden increase. Interestingly, the findings clearly indicated that optimum results were obtained with a down-swept FM sound. Specifically, when using an acrylic pole with a diameter of 2 mm, it was perceived by means of CF sounds at a distance of 62 cm from a subject, whereas use of a down-swept FM sound, resulted in the pole's perception at a greater distance of 92 cm.

In the case of a down-swept FM sound, the initial high-frequency sound emission makes it possible to detect even small obstacles, after which the emission changes to a low frequency. Although high sounds can easily be masked by low sounds, such a problem does not arise when using down-swept FM sounds as it is the high sound emissions that precede the lower ones. Based on the assumption that a bat's ears function in a way similar to that of humans, the advantage of down-swept FM sounds may therefore be the reason bats have developed the ability to utilize down-swept FM sounds for echolocation.

This section has offered two examples of mobility aids for the blind that were modeled on the echolocation used by bats. Although the results of this research have not yet to be put into practical use, it has become clear that the function of the auditory sense of bats and human beings is indeed similar. As such, there is a real possibility of developing superior mobility aids by further analysis of the echolocation mechanism of bats.

7.3   Obstacle Sense of the Blind

Even without the use of the mobility aids mentioned above, many blind people are nonetheless capable of perceiving obstacles to a limited degree. In contrast, most people with normal eyesight can barely walk when blindfolded or in the dark. The reason for such a difference is that the greater reliance on the auditory sense and/or nervous system of the blind has made that sense and/or nervous system particularly well developed. As a result, they are somehow able to perceive their surrounding environment well enough to avoid obstacles, even without groping their way or using any visual aids. This ability to sense one's surroundings without visual perception is known as the "obstacle sense."

Stated another way, in order to recognize their surroundings to the best degree possible, blind people seem to naturally compensate for their lack of sight by learning to make optimal use of their remaining senses. During this process of adaptation, however, it is the obstacle sense that plays a particularly important role. Specifically, such a means of perception is defined as "the ability to recognize and orient oneself to the existence of objects that are neither within one's reach nor seem to emit any particular sensory-based hints as to their location without the use of sight."

In instructional books for the blind, only a simple description is found regarding any method of training for developing this obstacle sense. In particular, although some manuals do mention that changes in sound can provide certain clues to the blind about their surroundings, no detailed description is offered. Therefore, no scientific or systematic methodology to develop the obstacle sense has of yet been established. Generally speaking, the blind are merely advised to learn from experience. Furthermore, for those who have already acquired such an obstacle sense, no methods exist to assist them in improving on their ability. In order to develop training methods for the obstacle sense and its practical use, a more detailed description is essential.

7.3.1 Conventional Hypotheses

In Diderot's book about the obstacle sense, titled "Letter of the Blind," he described it as a mysterious ability that enables those who have mastered it to not only perceive obstacles and their relative distance from them, but to also be able to distinguish such characteristics as the obstacles size, shape and material [19].

The various hypotheses regarding this obstacle sense can be divided into two basic theories, based on introspective reports by the blind. These two theories are known as the "auditory sense theory" and the "skin sense theory." The former is based on the concept that the existence of obstacles can actually be somehow *heard*, whereas the latter theory is based on the assertion that the blind *perceive* obstacle sensations on the face, forehead, and vertex or top of the head.

Regarding the latter theory, an experiment was once conducted that involved covering the entire body of blind subjects with a cloth in order to block any possible sensation felt by their skin. Despite this procedure, however, results proved that subjects were still able to perceive obstacles. In addition, a brief mention should be made of occult theories (or supernatural theories) regarding the obstacle sense. However, no clear empirical evidence has been found to support these theories, as research shows that blind subjects can still distinguish obstacles even when their surroundings do not contain any magnetic or electrical fields.

At Cornell University in the 1940s, a series of research projects was begun by Supa, Cotzin and Dallenbach in an attempt to verify the auditory sense theory [20, 21]. Their findings indicated that blind subjects were unable to perceive obstacles without the presence of some auditory hints, even when various other forms of non-auditory stimulation were introduced. In contrast, they also found that the obstacle sense still works even when other non-auditory forms of stimulation are blocked, thus indicating that the only crucial factor necessary for blind subjects to perceive obstacles is some form of auditory hint. This provided clear evidence that auditory stimulation is an essential clue to understanding the functioning of the obstacle sense.

After that time, the primary focus of much of the subsequent research was to investigate the obstacle sense phenomenon through psychological experiments rather than through analyzing the main factor—namely, auditory clues—responsible for the obstacle sense. In fact, most of that research was conducted without a full knowledge of acoustics. During the following 30 years, although it was clear that obstacle recognition was only possible by use of the auditory sense, very little was known regarding the actual mechanism involved. However, it should be possible to design glasses that could actually reinforce the obstacle sense or perhaps to formulate new methods of rehabilitative therapy resulting from knowledge obtained through a more thorough investigation of this seemingly mysterious ability.

7.3.2 Obstacle Sense While Walking

In cooperation with students at Japanese schools for the blind, Seki and the author applied acoustic knowledge to conduct research in an effort to more clearly determine how acoustic clues serve as an aid to obstacle perception [22]. In conducting our experiment, three completely blind people, aged between 17 and 19, were asked to walk toward a wall amid surrounding weak noise. As Fig. 7.14 shows, they stated that they could faintly hear the echoes from their own footsteps when the wall was at a distance of 5–6 m in front of or behind them. Therefore, just as in the case of bats, it is clear that the blind can in fact *hear* the existence of a wall by means of echolocation.

When the wall was about 3 m away, the subjects explained that they actually "felt" the existence of something. This feeling is referred to as a "first perception." When the wall was only 30–50 cm away, they stopped walking and claimed that they had the feeling that they might hit something. This sensation is termed the "final appraisal."

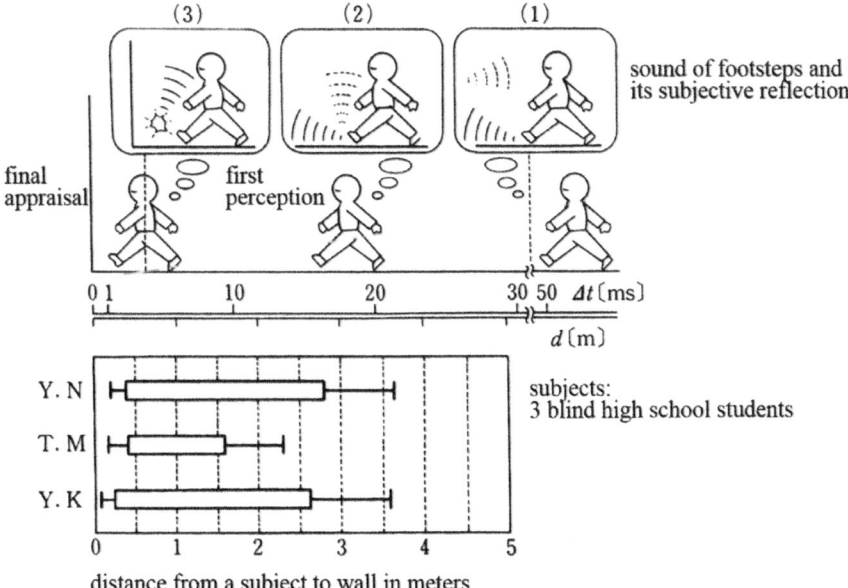

Fig. 7.14 Two steps of obstacle detection: "first perception" and "final appraisal" in walking toward a wall

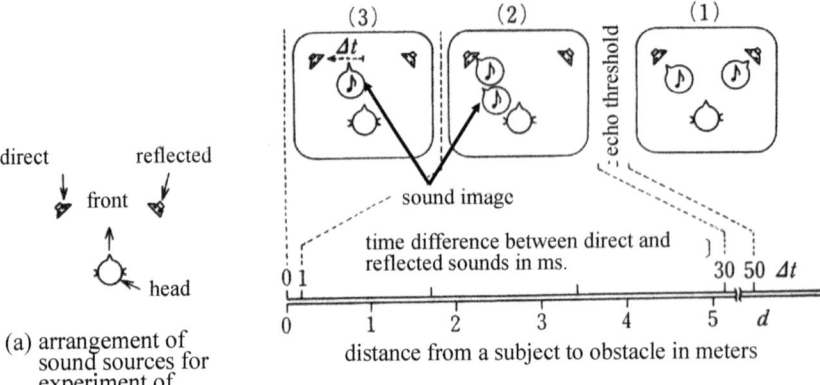

Fig. 7.15 a Arrangement of direct and the reflected sound sources, b change in lateralization as a function of time difference between two sounds

It appeared that some form of acoustic change occurred psychologically at a distance of between 30 and 50 cm from the wall. Although it was, at first, "an unknown," a new fact soon became evident; namely, that the "preceding effect" functions in such a way that sound that is reflected off the wall at a distance of 3 m is heard as a result of the direction of sound emanation from one's footsteps. Moreover, it is possible that this effect is actually combined with *the feeling that something is there.* Thus, the preceding effect could in fact be the same kind of phenomenon that people experience while listening to stereo music.

The top part of Fig. 7.15 schematically illustrates the preceding effect. When the time difference (Δt) is large between the left and the right speakers, e.g., more than 50 ms, the sounds that are temporally separated can be heard from both speakers. However, when Δt decreases, the latter sound gradually approaches the former sound until the latter sound is eventually heard from the direction of the former. It was anticipated that this preceding effect also occurs at "first perception."

Therefore, we conducted an experiment in which subjects were asked to point in the direction of the perceived sound while decreasing the delay time between the direct sounds from their footsteps and the reflected sounds (Fig. 7.16). The result shows that the reflected sounds start to be heard from a person's footsteps once the time difference reaches 10 ms. It is assumed that when the reflected sounds heard in front of a person descend to the feet, it evokes the first perception that something is there .

The results of our next investigation provided evidence that obstacles can even be detected with exceptional accuracy without auditory clues from footsteps. For this experiment, a carpet was spread over the floor to eliminate the possibility of hearing footsteps, as shown in of Fig. 7.17a. Moreover, during this state, with the use of white noise to simulate the various environmental noise emanating from the area behind the obstacles, blind subjects were asked to walk around several times in

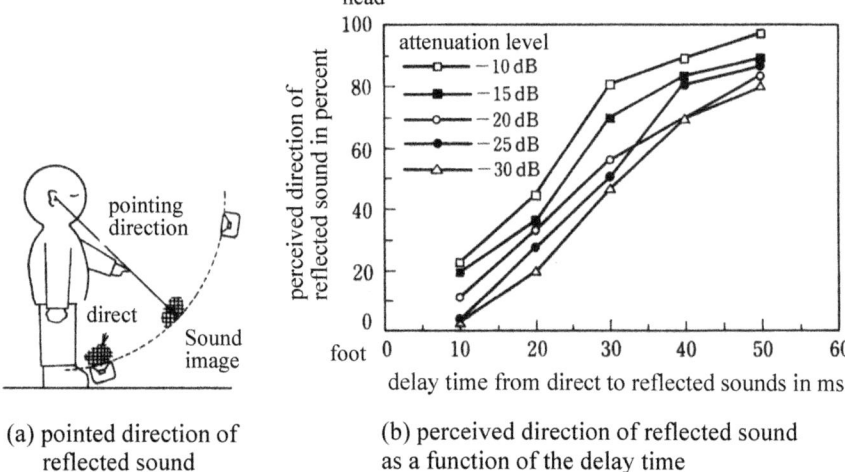

(a) pointed direction of reflected sound

(b) perceived direction of reflected sound as a function of the delay time

Fig. 7.16 Change in localization of reflected sound by preceding effect

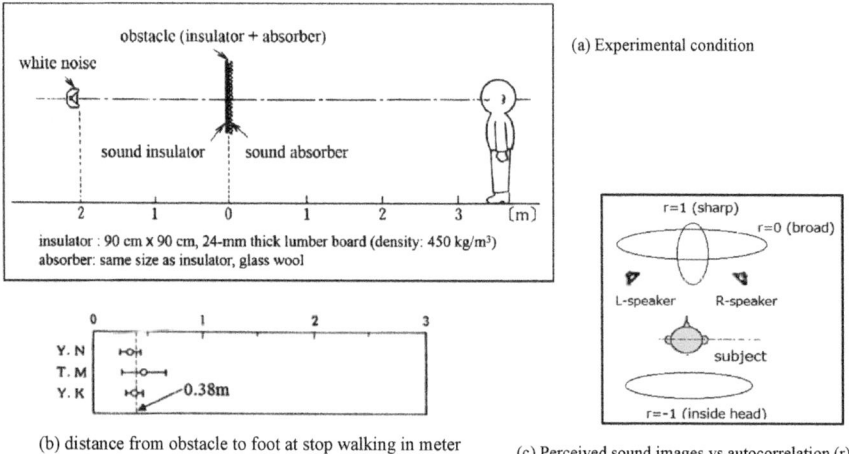

(b) distance from obstacle to foot at stop walking in meter

(c) Perceived sound images vs autocorrelation (r)

Fig. 7.17 a Experiment in obstacle sense, **b** its results in a sound-insulated condition. **c** Broadness of a sound image as a function of cross correlation between left and sides sounds

order to assess their ability in obstacle perception. Our results showed that all of the subjects clearly stopped walking at approximately 38 cm from an obstacle, even without the preceding effect of the sound of footsteps, as shown in the bottom left part of the Fig. 7.17b.

During this time, measurements were made of the change in sound pressure as well as the inter-aural correlation coefficient. The inter-aural correlation coefficient is used to ascertain the correlation between the sound waves that enter both ears, which is related to the spread of the sound image, as schematically shown in Fig. 7.17c.

The measurement result showed that the sound pressure gradually begins to drop at a distance of 1 m, while the inter-aural correlation coefficient suddenly falls at a distance of tens of centimeters. Therefore, the fact that a blind person stops at a distance of roughly 38 cm implies that the correlation coefficient is close to 0, indicating that a sound characterized by a weak sound pressure can suddenly be perceived to be widely spread out [23]. There is a possibility that this perception is related to the feeling of hitting something. Thus, it is hypothesized that the preceding effect is the cause of the first perception, whereas the sound pressure combined with a decrease in the inter-aural correlation coefficient is the cause of the final appraisal.

7.3.3 Obstacle Sense During a Resting State

Furthermore, from a preliminary experiment, we found that even when blind subjects were 50 cm from an obstacle (a 50 cm^2 lumber board) in a state of rest, they were able to use their auditory sense to differentiate between the presence and an absence of the obstacle by detecting certain changes in the sound field. However, when the same experiment was conducted using sighted subjects, they were unable to distinguish the object's presence by use of the auditory sense. As this experiment was conducted outdoors, there were various environmental noises. It is possible to hypothesize that the blind can feel and use the subtle changes caused by an obstacle's presence as a clue for detecting obstacles.

Therefore, in order to examine how the sound field changes depending on the presence of obstacles, we examined the change in the spectrum (acoustic transfer function) by using a dummy head in place of a blind person. The purpose of this experiment was to examine what sounds could be detected by a microphone fitted to the dummy head when a 90 cm^2, 24-mm thick lumber board was placed in front of it, while white noise was being presented from behind, as shown in the arrangement of the right figure of Fig. 7.18.

By examining the spectra, we could determine the number of dip changes that occurred depending on the distance to the obstacle. In the case that the phase of the noise and the reflected sound—which reaches the ears later—is reversed, a dip occurs in the spectrum, whereas when the phases coincide, a peak occurs in the spectrum. It is understood that the delay time Δt, which indicates the distance to the obstacle, determines the number of dip changes within the spectrum that is related to a sound's pitch.

The pitch of the sounds detected by the dummy head was higher near the obstacle, but lower when the obstacle was further away. This phenomenon is called "coloration" and plays an important role in the perception of sound fields, as shown in the left part of the figure. To verify that this hypothesis is, in fact, one of the main

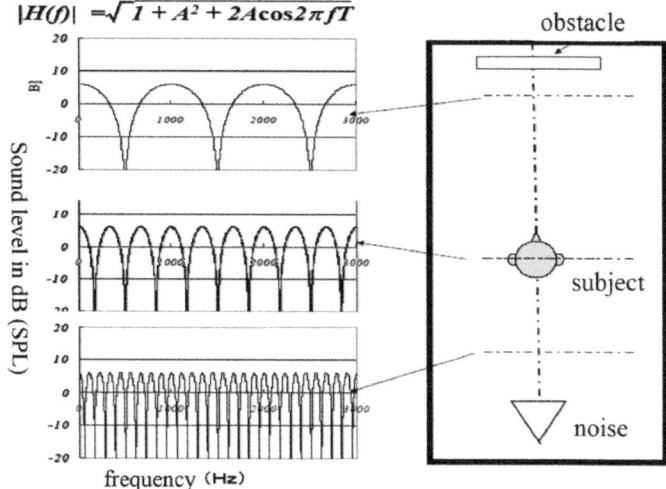

Fig. 7.18 Spectral changes caused by phase interference between direct sounds and reflected sounds from an obstacle

factors contributing to the obstacle sense, we conducted a simulation experiment. After having a blind person sit midway between two speakers, one speaker was made to produce direct sound while the other produced a sound with a delay of Δt. After hearing the resultant sound, the blind subject explained that he felt the presence of an obstacle. Furthermore, the blind subject answered that the larger the value of Δt, the further away the obstacle was perceived to exist.

7.3.4   The Role of Coloration Changes

The time delay between the two sound waves at the subject's ears forms a comb-like spectral pattern with dips, as shown in the left figure of Fig. 7.18. The frequency difference between the dips is the reciprocal of the time delay. Therefore, coloration is an important information cue in identifying the distance from the obstacle to the subject. The spectral amplitude of a synthesized sound wave consisting of direct and reflected sound wave $|H(f)|$ is expressed by:

$$|H(f)| = \sqrt{1 + A^2 + 2A \cos 2\pi fT} \qquad (7.1)$$

where f, A, and T represent the frequency, reflection coefficient for the obstacle, and arrival time interval between the direct and reflected sound, respectively. Here, the dip-to-dip interval in the spectrum is defined as the dip-frequency interval Δf_d, and the relation between Δf_d and the distance from obstacle d is:

$$d = c/2\Delta f_d \qquad\qquad (7.2)$$

where c is the sound velocity. Miura and the author determined the discrimination threshold of Δf_d corresponding to that of the obstacle distance d for both blind and sighted people, both with normal hearing. In addition, acoustic impressions regarding the change in Δf_d were checked from the experiments described below [24].

Equation (7.1) is the same as the frequency response of the comb filter. Thus, sound waves caused by the coloration can be generated by applying a comb filter to white noise that may simulate the surrounding noise in the real world. It has been reported that the comb-filtered noise called "ripple noise" has a matched pitch equal to the inverse of the time delay, and produced pitches similar to missing fundamental pitches [25].

In the experiment, subjects (six congenital visually impaired people aged 23–34 years old and six sighted people aged 23–42 years old, all with normal hearing) were asked and to answer whether two successive sounds with a different dip-to-dip interval were perceived to be the same by means of two alternative forced choices. The results showed no significant difference in discrimination between the blind and the sighted groups. From this result, as Δf_d is about $0.02 \times f$ as shown in solid line of Fig. 7.19, the discriminated distance Δd is also $0.02 \times d$. This means that the subjects could discriminate a ± 4 cm distance change when the obstacle was located at 2 m from the subjects and a ± 1 cm distance change at 1 m.

As to "impressions" elicited by the sounds with various dip-to-dip intervals, it was found that the sighted tended to focus mainly on quantitatively represented

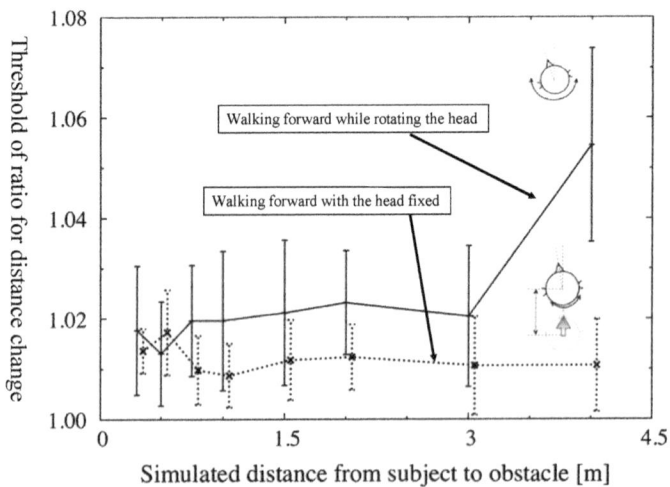

Fig. 7.19 Threshold of ratio (moving distance/distance from subject to obstacle) for distance change as a function of the distance from subject to obstacle. Comparison of that in the case of the head rotating and in the case of the head being fixed

changes such as pitch and loudness of the sounds, while the blind were inclined to focus not only on quantitative sound changes, but also on qualitative impressions such as "powerful-unsatisfactory," "annoying-quiet" and "beautiful-ugly." in the sound changes.

As it was assumed that qualitative impressions in addition to qualitative impressions are related to the perceived distances of the obstacles in the case of the blind subjects, the ability to perceive the distance from the obstacle should be superior for the sighted subjects.

In daily life, as blind people as well as sighted people search for and find obstacles while walking or working, their ears move to the front, back, left and right depending on the movements of their head and body trunk. It is assumed that the movements would improve the discrimination ability of the distance from the obstacles. Miura and the author continued with the research and compared the discrimination ability in the case when the subjects' heads were fixed and in the case when their heads rotated while walking. As shown in the broken line of Fig. 7.19, the results show that the ratio of the discriminated distance Δd to the distance d from the obstacles clearly decreased, that is, the distance discrimination ability increased when the subjects walked while their heads were rotating.

Based on these results, Seki proposed new concepts for blind mobility aids as well as an obstacle-sense training method by constructing an acoustic virtual reality system, as shown in Fig. 7.20. In his system, various sound effects such as attenuation, delay, diffraction, insulation, the Doppler effect, and a head-related transfer function (HRTF) are incorporated in a personal computer to enable users to recognize virtual environments. The users can perceive virtual obstacles and avoid them inside the virtual environment while walking, solely by acoustic clues presented through headphones. It is thus proved that the system is indeed useful for training the obstacle sense of the blind without any special equipment.

This series of experiments demonstrates that what the blind refer to as "*a sign or feeling*" that an obstacle exists is, in fact, caused by subtle changes within a sound spectrum, especially those due to coloration. Such a phenomenon as the obstacle sense is due to the "plasticity" of the human brain; namely, it's distinct ability to substitute a certain function for another dormant function when sight has been lost

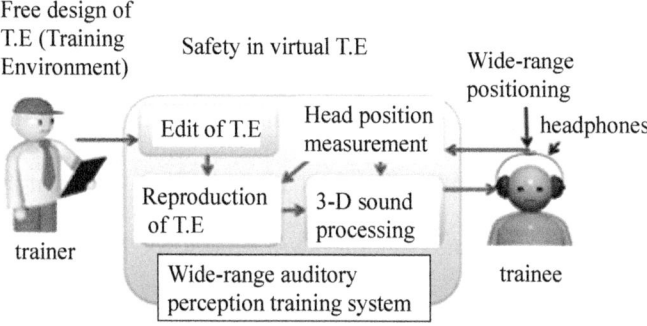

Fig. 7.20 Practical training system for spatial perception by using a PC and headphones

or impaired. This example of plasticity involving the function of the auditory sense suggests that there may be many more similar undeveloped abilities that humans possess that have yet to be discovered.

Provided that such an ability does in fact exist, it is no mere fantasy to emphasize the importance of auditory clues such as coloration, the preceding effect and the inter-aural correlation coefficient so that obstacles can be more clearly perceived or emphasized. Additionally, by surrounding a person with speakers and controlling the sound source emanating from each speaker, it could also be possible to evoke feelings that various objects are present. Such an apparatus might constitute the ultimate development in the virtual reality of sound [26].

7.4 Sound Localization and Balance

Somewhat putting aside research on assistive technology for the blind, this section will explain and offer examples of how virtual reality sound technology is used in assistive technology related to the sense of balance.

When considering the kind of virtual reality technology that has started to be applied to a variety of fields, it is clear that most of it makes use of the visual sense to provide information related to feelings such as location, movement and turning. In contrast, since auditory information is processed unconsciously, it is possible for people to perceive movement or rotation without experiencing any discomfort. A great deal of research has been conducted to enable a person to experience a greater feeling of immersion in a virtual environment by presenting surroundings to the user through the combination of body movement and auditory information. However, contrary to what one would expect, while a person or sound image is moving, the relation between the sense of balance and the sound image localization is still unknown.

7.4.1 Mechanism of the Balance Sense

The sense of balance arises from several integrated senses, which include the vestibular sense, the visual sense and the somatosense. As shown in Fig. 7.21, the vestibular sense is composed of the semicircular canal (vertical semicircular and horizontal semicircular), and the otolith (sacculus maculae, utriclus maculae). Both the semicircular canal and the otolith are connected with lymphatic fluid via the cochlea. The receptive cells of both organs are hair cells, just like in the auditory sense. Within the head, it is assumed that the semicircular canal mainly receives angular acceleration, while the otolith receives straight acceleration.

Therefore, when the ear is filled with water and the inner ear is cooled, temporary dizziness occurs because pressure increases on the lymphatic fluid in the semicircular canal and other organs. As a result of this process, a person's balance

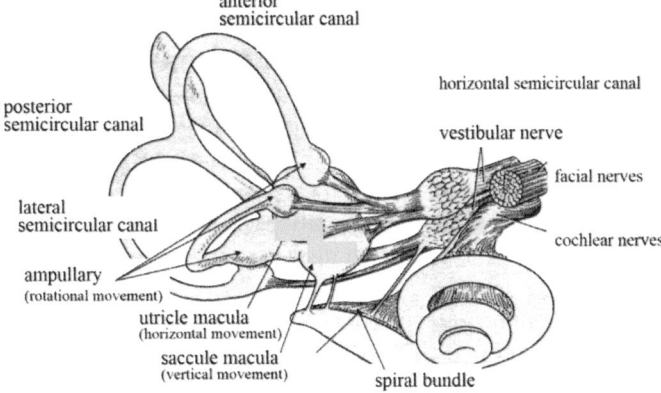

Fig. 7.21 Schematic representation of the equilibrium organ

function can be examined by using a method known as a "caloric test." As the semicircular canal, the otolith and the cochlea are also closely connected at the distal region, both defective hearing and dizziness can occur simultaneously.

7.4.2   Body Posture Control by Stationary Sound [27]

The author would like to describe how research on sound localization in the design of virtual reality serves to control one's sense of balance; this includes such factors such as how to decrease any feeling of dizziness or how to enable a person to keep herself or himself in an upright position. A description will now be given as to how the sense of sound localization changes according to the presentation of sounds when transient stimulation has been applied to the sense of balance. For this purpose, as shown in Fig. 7.22, a situation was created in which the sense of balance

Fig. 7.22 Experimental setup for an equilibrium test —the influence of sound stimulation on optical-flow phenomena

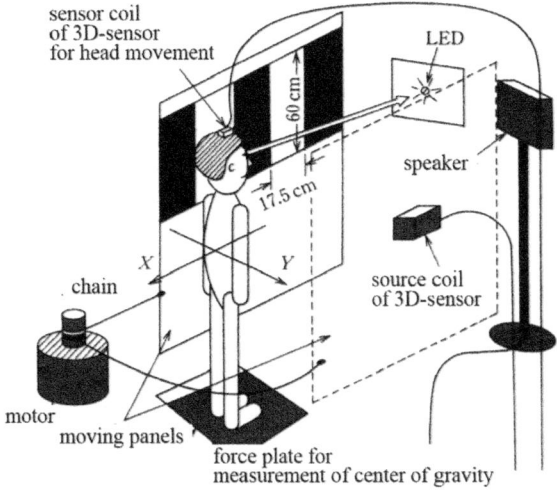

was influenced when panels with a striped pattern were moved in the area of sight known as "far peripheral vision."

As the figure shows, the panels that had been placed on both sides of each subject's far peripheral vision were moved backward at a speed of 17.5 mm/s. A speaker was placed 30° to the right of the subjects to generate white noise [52 dB (A)]. For this experiment, the subjects (6 males with normal hearing) were instructed to stand upright on the force plate and to look squarely in front of them at a red LED.

First, when the striped pattern was slowly moved in the area of far peripheral vision, an "optical flow" was perceived; that is to say, subjects felt like they were moving in the opposite direction to the striped pattern. When this experiment was conducted with the subjects standing on an instrument to measure the shift or disturbance of their center of gravity (COG), the COG was found to move toward the direction of the striped pattern's movement, as if to compensate for the confusion in the subject's sense of balance as shown in Fig. 7.23a.

Next, whether the postural inclination is influenced by the existence of sound was investigated. From the result, it is clear that the degree of inclination of the subject's body (toward the same direction as the optical flow) decreased as a result of the presence of a sound source. Figure 7.23b shows the COG obtained after the addition of a stationary sound source. Such findings make it obvious that the existence of a sound source decreases with a back-and-forth motion. Additionally, a review of subjective impressions gave added support to the conclusion that with the

Fig. 7.23 Change in center of gravity by optical stimulation—with (**b**) without (**a**) sound stimulation

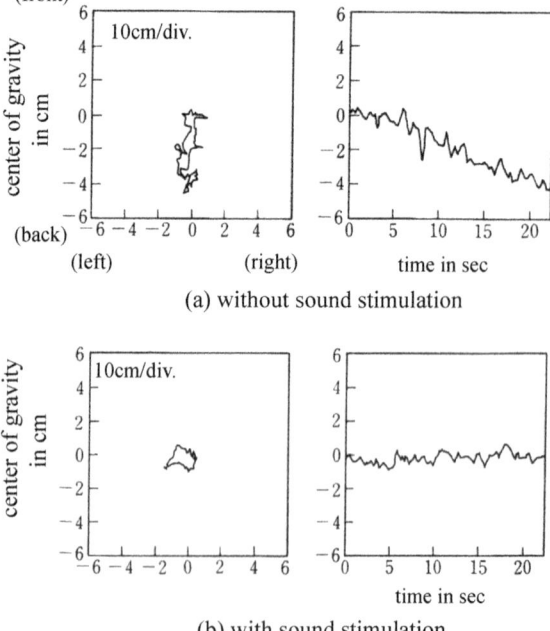

influence of a stationary sound source, subjects experienced a decrease in the sensation of being pulled backward, as compared to the experiment that only utilized optical flow. It should be noted that such a phenomenon was not observed when headphones were used to hear the sound.

Within the confines of the experimental results, it can be said that the postural disturbance or confusion can be decreased through the presentation of a stationary sound source that serves to buffer the influence of optical flow on a person's sense of balance. Judging from this, there is a possibility of correcting information received by the auditory sense in situations when a person's visual sense gives rise to a disturbance in one's sense of balance.

Therefore, in the case of virtual reality, when the operator's sense of balance is somehow confused, it might be helpful to apply a method of adding auditory information to the visual information. Moreover, it would certainly prove worthwhile to make an effort to find useful applications of sound localization to rehabilitation therapy for people suffering from functional disorders of balance.

7.4.3 Interaction Between Balance Function and Sound

To gain a more realistic feeling of immersion in virtual reality, in addition to the visual sense, information must also be provided to both the auditory sense as well as the sense of balance. We first investigated how sound image localization is influenced by rotating oneself. The setup used in this investigation is shown in Fig. 7.24. In order to examine the impact of this influence, the angular velocity (ω_h) of the subject's rotation was altered using a rotating chair. Sound with an angular velocity of ω_s was presented during the rotary motion. For this experiment, the subjects were 7 males with normal hearing, aged from 23 to 30. Each of the subjects wore an eye mask and sat on a rotating chair in a dark anechoic chamber. Sixteen speakers, each 12 cm in diameter, were used as the sound source and were placed in a circle 1.5 m away from the subjects on a horizontal plane. White noise of [56 dB (A)] was used as the sound source signal.

Before beginning this experiment, each subject's vestibular nerve was stimulated by the use of a rotating chair in order to cause temporary dizziness. Without going into the details of this experiment, suffice it to say that our results indicated that subjects perceived the counterclockwise rotation of the sound when the subject's rotation velocity ω_h was much larger than the sound source rotation velocity ω_s, as shown in the bottom part of the figure. In short, the rotation clearly resulted in a greater shift or gap in the subject's localization to the sound image.

Conversely, we investigated how the existence of a moving sound source influences the sense of balance when a person perceives a feeling of rotation. When the rotating sound was presented while rotating the subject, the subject were asked to indicate the subjective rotating speed in 10 steps. The findings indicated that each subject's own rotation was perceived to be slower than the actual speed when the sound rotated in opposite direction of the subject's rotation.

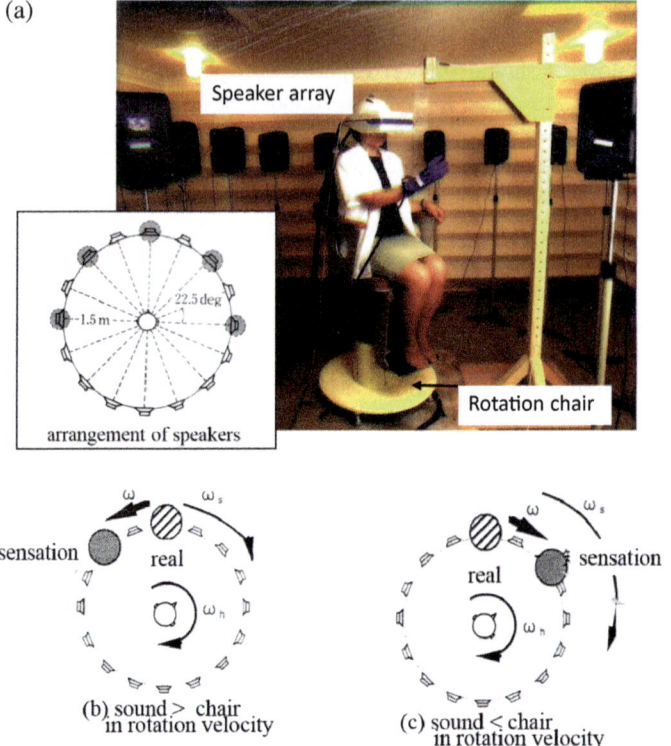

Fig. 7.24 *Top* Experimental setup for localization test, arrangement of rotating chair and speaker array. *Bottom* Changes in sound localization and sensation of rotation due to difference between sound and chair in rotational velocity

Based on our observations, it is assumed that in the human perception of rotary motion, sound can serve as a standard frame of reference whose center is oneself. Therefore, it is possible to say that the feeling of change in the rotational speed of human beings can be either overestimated or underestimated by changing the relative angular velocity between the person and a sound source. These observations provide hints as to how researchers could make the best use of a moving sound source in order to decrease functional disorders of balance. In the field of acoustic assistive technology, it is important to clarify this connection between the sense of balance and sound localization.

7.4.4   Balance Control by Sound Localization [28]

Tanaka and his co-researchers investigated how moving auditory stimuli could affect the standing balance in healthy adults of different ages. The participants of the research were 12 healthy volunteers, who were divided into two categories: the young group (mean = 21.9 years) and the elderly group (mean = 68.9 years). The instrument used for the evaluation of standing balance was a force plate for measuring body sway parameters. The toe pressure was measured using the F-scan Tactile Sensor System. The moving auditory stimulus produced a white-noise sound and binaural cue using the Beachtron Affordable 3D audio system.

The moving auditory stimulus conditions were employed by having the sound come from the right to left or vice versa at the height of the participant's ears. Participants were asked to stand on the force plate in the Romberg position for 20 s with either eyes opened or eyes closed for analyzing the effect of visual input. Simultaneously, all participants tried to remain in the standing position with and without auditory stimulation from the headphones. In addition, the variables of body sway were measured under four conditions for analyzing the effect of decreased tactile sensation of toes and feet soles: standing on the normal surface (NS) or soft surface (SS) with and without auditory stimulation. The participants were asked to stand in a total of eight conditions.

The results showed that the lateral body sway of the elderly group was more influenced than that of the young group by the lateral moving auditory stimulation as shown in Fig. 7.25. An analysis of toe pressure indicated that all participants used their left feet more than their right feet to maintain balance. Moreover, the

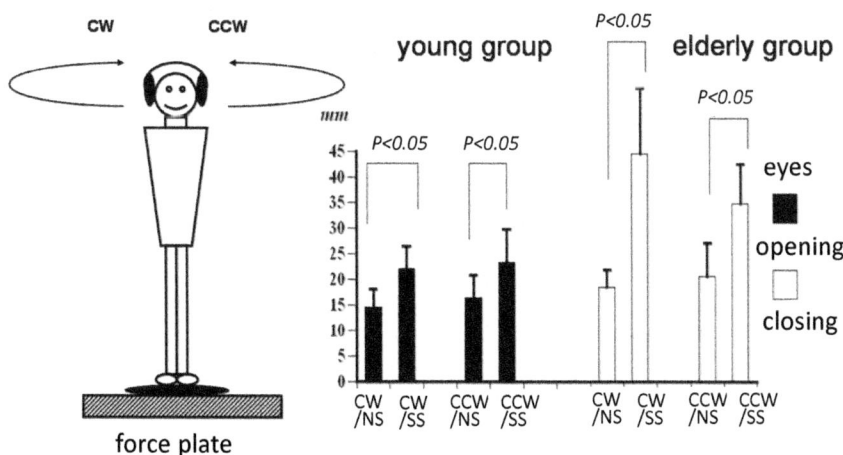

Fig. 7.25 *Left* Experimental setup using sound localization stimulation, *right* comparison of the young with the elderly

elderly had the tendency to be stabilized mainly by use of their heels. The young group were mainly stabilized by their toes. The results suggest that the elderly may need a more appropriate stimulus of tactile and auditory sense as a feedback system than the young for maintaining and control of their standing postures.

7.5   The Future of Assisting Technology for Seeing

Recently, thanks to steady progress in technology, new interfaces have been appearing that have attracted considerable interest from the blind. Actually, research on developing practical interfaces between the blind and computers is now actively being conducted as society gradually moves towards a new era of multimedia. However, in researching and developing assistive technology for the blind, most researchers have been limited from a technological point of view. This is because it is difficult for researchers working in the area of rehabilitation and education of disabled children to gain cooperation or understanding from other professionals who are actually working in the field. In the overall area of assistive technology, including assistive devices for the blind, the key to the promotion and widespread use of such technology is the completion of appropriate training manuals and clearly established training methods for the use of such devices coupled with the development of a reliable supply and maintenance system.

Although there is a tendency in the field of sound-assistive technology to place undue emphasis on advanced technology to assist people with auditory and vocal defects, assistive technology could also prove useful in alleviating other symptoms, such as dizziness and motion sickness, which at present seem unrelated to sound. Therefore, the time is ripe for researchers to make greater efforts in finding new applications for assistive technology in a broader area that may offer great potential for enhancing the lives of other people whose symptoms do not fall within the traditional, but narrow, categories of auditory or vocal disability.

The rapidly spreading use of car navigation devices, mobile phones and robots equipped with vocal sound output offers great potential for new and more practical applications of assistive tools for the blind. Unfortunately, in the case of the congenitally visual impaired, navigation systems are not applicable in their present states as it is difficult for such people to develop the basic concept of maps. In marked contrast, it is assumed that the acquired visually impaired will not require much training in order to learn how to operate assistive devices that already make full use of the latest technology.

It is important that technologists actively investigate the extent to which such advanced technology can be applied to assist the impaired. In conclusion, the time has now come for people who deal directly with the visually impaired on a regular basis to put aside any misgivings or reluctance they may have regarding advanced technology, and come to realize its importance and future potential in improving the quality of life for countless disabled people throughout the world.

References

1. A. Koide, C. Asakawa, N. Suzuki, Research on computer aids for the visually disabled. in *Proceedings of the IISF/ACM Japan International Symposium: Computers as Our Better Partners*, Tokyo pp. 66–69 (1994) [in Japanese]
2. urzweil 1000, Kurzweil Educational Systems, Inc. http://www.kurzweiledu.com/(2016.9.20)
3. M. Shinohara, Vocal character reader for persons with disabled sight. J. Acoust. Soc. Jpn. **43**(5), 336–343 (1987). [in Japanese]
4. T. Watanabe, S. Okada, T. Ifukube, Development of GUI screen reader for blind persons. J-81-D-II(1), 137–145 (1998)
5. C. Asakawa, H. Takagi, S. Ino, T. Ifukube, TAJODA: Proposed tactile and jog-dial Interface for the blind. IEICE Trans. Info. Syst. E87-D(6), 1045–1014 (2004)
6. C. Asakawa, H. Takagi, S. Ino, T. Ifukube, The optimal and maximum listening rates in presenting speech information to the blind. Proc. ICAD **7**(1), 105–111 (2005). [in Japanese]
7. M.P. Beddos, C. Suen, Evaluation and a method of presentation of the sound output from the Lexiphone-A reading machine for the blind. IEEE Trans. BME-18, 85–91 (1971)
8. K. Ito, Y. Yonezawa, An application of sound localization effect to pattern display. Trans. Inst. Electron. Commun. Eng. C, J51-C-12, 753–760 (1978) [in Japanese]
9. T. Ifukube, I. Ushioda, C. Yoshimoto, Analyses of multi-sound localization for substitutional cognition for the blind. Technical Report Hearing Research of Journal of the Acoustical Society Japan H-46-3, 1–8 (1977) [in Japanese]
10. J. Blauert, Spatial Hearing, S. Hirzel Verlag. Japanese edition: Kuhkan Onkyo (Spatial Hearing), (translated by Morimoto M, 1986), (Kajima Institute Publishing Co., Ltd 1973)
11. H. Wallach, On sound localizaton. J. Acoust. Soc. Am. **10**, 270–274 (1949)
12. M.B. Gardner, R.S. Gardner, Problem of localization in the median plane: effect of pinnae cavity occlusion. J. Acoust. Soc. Am. **53**, 400–408 (1973)
13. T. Ifukube, C. Yoshimoto, Character recognition by auditory latelarization. Technical Report of Hearing Research in Journal of the Acoustical Society ET78-13, 9–12 (1979) [in Japanese]
14. T. Sasaki, Mobility aids for blind using ultrasound. J. Acoust. Soc. Jpn. **43**(5), 344–348 (1987). [in Japanese]
15. J.M. Benjamin Jr., A.A. Nazir, An improved laser cane for the blind. in *Proceedings of SPIE 0040, Quantitative Imagery in the Biomedical Sciences II*, 101 (1974)
16. T. Sasaki, Electronics for walking support of the visually handicapped. BME, Biomed. Eng. **7**(7), 15–20 (1993). [in Japanese]
17. L. kay, Ultrasonic spectacles for the blind. in *Proceeding of International Conference on Sensory Devices for the Blind*, St. Dunstans (1966)
18. T. Ifukube, T. Sasaki, C. Peng, A blind mobility aid modeled after echolocation of bats. IEEE Trans. BME **38**(5), 461–465 (1991)
19. D. Diderot, Letter on the Blind (1749)
20. M. Supa, M. Cotzin, K.M. Dallenbach, Facial vision: the perception of obstacles by the blind. Am. J. Psychol. **57**, 133–187 (1944)
21. M. Cotzin, K.M. Dallenbach, Facial vision: The role of pitch and loudness in the perception of obstacles by the blind. Am. J. Psychol. **63**, 485–515 (1950)
22. Y. Seki, T. Ifukube, Y. Tanaka, Relation between the reflected sound localization and the obstacle sense of the blind. J. Acoust. Soc. Jpn **50**(4), 289–295 (1994)
23. K. Seki, T. Ifukube, Y. Tanaka, The influence of sound insulation effect on the obstacle sense of the blind. J. Acoust. Soc. Jpn **50**(5), 382–385 (1994)
24. T. Miura, T. Muraoka, T. Ifukube, Comparison of obstacle sense ability between the blind and the sighted: a basic psychophysical study for designs of acoustic assistivedevices. Acoust. Sci. Tech. **31**(2), 137–147 (2010)
25. S. Greenberga, J.T. Marsh, W.S. Brown, J.C. Smith, Neural temporal coding of low pitch. I. Human frequency-following responses to complex tones. Elsevier. Hear. Res. **25**(2–3), 91–114 (1987)

26. T. Ifukube, Artificial reality based on biomedical engineering: as an example case of sensory substitute studies. J. Inst. Telev. Eng. Jpn. **46**(6), 718–726 (1992) [in Japanese]
27. H. Nara, S. Ino, Y. Onda, T. Ifukube, A basic study on postural control using of moving sound image stimulation. Trans. Virtual Reality Soc. Jpn. **5**(3), 1013–1018 (2000)
28. T. Tanaka, S. Kojima, H. Takeda, S. Ino, T. Ifukube, The influence of moving auditory stimuli on standing balance in healthy adults with aging. Ergonomics **44**(15), 1403–1412 (2001)

Index

A
Absolute refractory period, 81
Accessory auricles, 214
Acquired deaf, 112
Acquired hearing disability, 150, 151
Action support, 164
Acute tympanitis, 103
Afferent nerve, 20, 78, 89
Affricative, 30
Ainu people, 57
Air-conduction hearing level, 44
Akira Ifukube, 73
Amplitude fluctuation, 184
Angular gyrus, 38
Angular velocity, 229
Antibiotics, 82
Aphasia, 38
Arcuate fasciculus, 38
Articulation, 148
Articulation disorders, 169
Artificial
    electro-larynx, 169
    larynx, 25
    middle ears, 43, 51
    neural networks, 151
    organs, 1
    retina, 77, 105
    vision, 105
Artificial intelligence (AI), 151
Auditory brainstem implants, 77
    dysphonia, 193
    localization substitutes, 111
    ossicle, 15
    primary neurons, 20
    substitute, 111
Auricles, 13, 208
Autocorrelation function, 60
Automatic gain control (AGC), 55
Average hearing level, 45

B
Backward masking, 120
Barrier-free design, 3
Basilar membrane, 15
Bats, 199
Bayes' theorem, 150
Békésy, 16
Bernoulli's principle, 25
Big data, 159
Bilabial consonants, 189
Bilateral directional characteristics, 187
Binaural headphone, 206
Bipolar cells, 106
Bipolar method, 88
Blindfolded, 217
Bone-conductive hearing level, 44
Bottom-up, 159
Braille, 112
Brain machine interface (BMI), 89
Brain plasticity, 89
Breath sounds, 26
Broca aphasia, 192
Broca's area, 38
Broker's area, 23
Bronchial tubes, 175

C
Captioning systems, 145
Care robotics, 6
Categorization, 148
Category decision test, 30
CELP, 30
Center of gravity (COG), 228
Central nerve deafness, 43
Central nervous system (CNS), 20
CF-FM bats, 213
Characteristic frequency, 21
Chinese characters, 154
Clark, 96

Printed by Printforce, the Netherlands